Disaster Resilience and Green Growth

Over the years, the relationship between environment and disasters has received significant attention. This is largely due to the emerging recognition that environmental changes - climate change, land-use and natural resource degradation make communities more vulnerable to disaster impacts. There is a need to break this nexus through environment based and sustainability inclusive interventions. Science – technology and economic measures for disaster risk management, hence, need to adapt more integrated approaches for infrastructure and social resilience. Environmental and anthropogenic factors are key contributors to hazard, risk, and vulnerability and, therefore, should be an important part of determining risk-management solutions.

Green growth approaches have been developed by emphasizing sustainability inclusion and utilizing the benefits of science-technology interventions along policy-practice linkages with circular economy and resource efficiency. Such approaches recognize the perils of traditional material-oriented economy growth models that tend to exploit natural resources, contribute to climate change, and exacerbate disaster vulnerabilities, Green growth integrated approaches are rapidly becoming as preferred investment avenue for mitigating climate change and disaster risks and for enhancing resilience. This includes ecosystem-based and nature-based solutions with potential to contribute to the resilience of infrastructure, urban, rural and peri-urban systems, livelihoods, water, and health. They can lead to food security and can further promote people-centric approaches.

Some of the synergistic outcomes of green growth approaches include disaster risk reduction, climate change mitigation and adaptation, resilient livelihoods, cities, businesses and industry. The disaster risk reduction and resilience outcome of green growth approaches deserve special attention, both for the academic and policy communities. Scholars and professionals across the domains of DRR, CCA, and green growth are in need of publications that fulfill their knowledge needs concerning the disaster resilience outcomes of green growth approaches. Keeping the above background in view, the book series offers comprehensive coverage combining the domains of environment, natural resources, engineering, management and policy studies for addressing disaster risk and resilience in the green growth context in an integrated and holistic manner. The book series covers a range of themes that highlight the synergistic outcomes of green growth approaches.

The book series aims to bring out the latest research, approaches, and perspectives for disaster risk reduction along with highlighting the outcomes of green growth approaches and including Science-technology-research-policy-practice interface, from both developed and developing parts of the world under one umbrella. The series aims to involve renowned experts and academicians as volume-editors and authors from all the regions of the world. It is curated and developed by authoritative institutions and experts to serve global readership on this theme.

Nerma Omićević • Bojana Bojanić Obad Šćitaroci

The Urban Rehabilitation of Post-Disaster Scapes

Springer

Nerma Omićević
Architecture Program
International University of Sarajevo,
Faculty of Engineering and Natural
Sciences
Sarajevo, Bosnia and Herzegovina

Bojana Bojanić Obad Šćitaroci
Department of Urban Planning, Spatial Planning
and Landscape Architecture
University of Zagreb, Faculty of Architecture
Zagreb, Croatia

ISSN 2662-4885 ISSN 2662-4893 (electronic)
Disaster Resilience and Green Growth
ISBN 978-981-19-9507-1 ISBN 978-981-19-9505-7 (eBook)
https://doi.org/10.1007/978-981-19-9505-7

This Springer imprint is published by the registered company Springer Nature Singapore Pte Ltd.
The registered company address is: 152 Beach Road, #21-01/04 Gateway East, Singapore 189721, Singapore

Preface

This book provides the first extensive examination and analysis of the use of the urbanscape during the disaster process, by connecting its elements throughout disaster phases: the pre-disaster phase, consisting of reduction in form of prevention and mitigation; the disaster event phase, consisting of the disaster impact followed by the disaster effects; and the post-disaster phase, consisting of the post-disaster recovery.

The scope of this research is primarily concerned with understanding how the urbanscape is used during the disaster process, while addressing the main questions: How is the urbanscape network important in every phase of the disaster process? How each of the urbanscape elements embodies a different role during the disaster process? We seek to answer these questions with the aim to reveal the importance of the urbanscape, during the disaster process, based on its use and the interconnectedness of its elements between the disaster phases and to determine the urban rehabilitation models of post-disaster cities.

The urbanscape in this paper is defined by the open public spaces within the city that include the natural and the built environment of the city. By identifying and understanding the quality of the open public spaces that form the urbanscape in the pre-disaster phase or the every-day life, open public spaces can reduce the impact of the disaster during the disaster event phase and influence the post-disaster rehabilitation during the post-disaster phase. This can be achieved by increasing disaster awareness and through the urbanscape perception and the use of its open public spaces in everyday activities. The ultimate aim is to ensure that disaster awareness and perception are being developed and implemented as of today and not from tomorrow.

The research includes the analysis of 18 disaster case studies in the world, of which 12 case studies are within the natural and 6 case studies within the man-made disaster category. The criteria for choosing these examples are based on the division into natural and man-made disaster subcategory and the period of their occurrence, from 1991 to 2021. The main purpose of this comparative analysis is to reveal the use of the urbanscape during the disaster process, through the role of its open public spaces during each disaster phase, in both natural and man-made disaster categories.

The comparative analysis of the use of the urbanscape during the disaster process within the selected disaster case studies will result in recognizing the possible post-disaster urban rehabilitation models. The post-disaster urban rehabilitation models are based on the criteria of the urban rehabilitation program, plans and strategies, used in the post-disaster recovery process, of both natural and man-made disasters.

The primary audience for this work are researchers in academia and students with a secondary audience of experts involved in the disaster management process, together with urban planners and architects.

Sarajevo, Bosnia and Herzegovina — Nerma Omićević
Zagreb, Croatia — Bojana Bojanić Obad Šćitaroci

Contents

About the Editors

Nerma Omićević is a professor at the Architecture Program within the Faculty of Engineering and Natural Sciences of the International University of Sarajevo. From 2011 until 2019, she worked at the position of Senior Assistant and since 2019 she has been appointed to the position of Assistant Professor. She lectured various courses within the program, supervised students' design projects, and contributed to many administrative duties on the program, faculty, and institutional level. She is the author/coauthor of scientific articles and diverse design projects in the field of architecture and landscape design.

Bojana Bojanić Obad Šćitaroci is a professor at the Faculty of Architecture of the University of Zagreb, the Department of Town Planning, Regional Planning, and Landscape Architecture. Since 1991, she has worked as an independent scientific researcher, planning engineer, and town planner. She is the author/coauthor of books, scientific articles, scientific studies, town plans, and studies and designs from the field of garden and landscape architecture. She has actively participated in scientific research projects at home and abroad in the field of landscape architecture and urbanism. She has taken part in national and international scientific-expert conferences (Zagreb, Stari Grad on the island of Hvar, Poreč, Varaždin, Palmanova, Udine, Vienna, Shanghai, Venice, Naples, Bari, Barcelona, Athens, Belgrade, Albena, Copenhagen, Lisbon, Pescara) with topics in the field of protection of traditional architecture, urbanism, and landscape architecture.

Chapter 1
Introduction

Abstract The aim of this research is to discover the significance of the urbanscape during the disaster process, by using it and connecting its elements throughout disaster phases. The urbanscape affected by natural disasters requires restoration of its physical structure—recovery, while the war-torn urbanscapes also require the recovery—redefinition of identity, and reinterpretation and discovery of its meaning in the post-war rehabilitation process. This research provides the first comprehensive overview and analysis of the use of the urbanscape during the disaster response process to set the urban rehabilitation models of cities exposed to disasters. Understanding the use of the urbanscape during the disaster process contributes to the establishment of the criteria for re-redefining the urbanscape identity. It is expected that these results can have an important scientific contribution by defining a part of the criteria for future urban planning.

Keywords Urbanscape · Identity · Rehabilitation · Natural disasters · Man-made disasters · Disaster impact · Post-disaster recovery · Redefined identity

1.1 Introduction

The opposing union of landscape and urbanism is mainly reflected through their mutual interconnectedness that derives from placing the landscape within the city and extending the city into the surrounding landscape. Due to a massive population growth, the traditional clear separation of these design professions evolved into a mutual response to the demanding forces of change, by contesting the term urban landscape or urbanscape. Urbanscapes refer not only to the clear structure of the built and unbuilt environment within the city, but they are also defined through the system of elements that occupy all open public spaces within the city. The dualistic dimension of the term urbanscape does not imply only the assimilation of the contested terms landscape and urbanism, but is also reflected through the dualistic role of their elements in the everyday urban life and during the impact of change.

The changes in today's urbanscapes are mainly influenced by the dynamic interaction of the cultural and natural shaping forces that led to the alteration of

N. Omićević, B. Bojanić Obad Šćitaroci, *The Urban Rehabilitation of Post-Disaster Scapes*, Disaster Resilience and Green Growth,
https://doi.org/10.1007/978-981-19-9505-7_1

both, the landscape and urbanism. These changes are mainly reflected through the loss of diversity and coherence in the spatial sense that eventually leads to changes of the urbanscape identity.[1]

The main forces of change through history were accessibility, urbanization and the unpredictable factor, the disasters.[2] Today's urbanscape is mainly altered due to the increasing impact of disasters. The United Nations International Strategy for Disaster Reduction defines disaster as a "serious disruption of the functioning of a community or a society involving widespread human, material, economic or environmental losses and impacts, which exceeds the ability of the affected community or society to cope using its own resources".[3] According to the main disaster classification, they can be divided into two main categories, disasters caused by natural forces and man-made disasters. Natural disasters are caused by natural forces, such as earthquakes, volcanic eruptions, floods, etc., while the man-made disasters are caused by non-intentional and intentional human actions, such as terrorism and conflicts of war. Regardless of their categorization, disasters are defined through their risk and vulnerability leading to short-term and long-term impact on the society and the urban fabric.

The city is defined through its urban fabric by "the topography and the organization of open spaces and the cohesive character of the built-up areas within the city",[4] whose interrelations influence the formation of the urbanscape or transformation in the post-disaster context. The urbanscape in this paper is defined by the open public spaces within the city that include the natural and the built environment of the city. The built environment, through its open public places (streets, squares, bridges, etc.) and urban structure, enables urban interaction and gathering, while the natural environment that enhances the green spaces (parks, greenways, rivers, surrounding hillside) provides sustainability.[5,6]

The use of the urbanscape and its network contributes to the quality of the everyday life and at the same time plays an important role in the disaster process.[7] The disaster process is composed of three independent phases that combine into one event: the pre-disaster phase, consisting of reduction in form of prevention and mitigation; the disaster event phase, consisting of the disaster impact followed by the disaster effects; and the post-disaster phase, consisting of the post-disaster recovery.[8]

The International Strategy for Disaster Reduction (ISDR) defines post-disaster recovery as "decisions and actions taken after a disaster with a view to restoring and

[1] Antrop (2005:1)

[2] Antrop (2005:5)

[3] UNISDR (2009:9)

[4] Aga (2010:61)

[5] Omićević and Obad Šćitaroci (2018)

[6] Saxena (2016)

[7] Omićević and Obad Šćitaroci (2018)

[8] NIDM:2

improving the pre-disaster living conditions of the stricken community, while encouraging and facilitating necessary adjustments to reduce disaster risk".[9] As a link between the disaster response and the long-term sustainable development, the post-disaster recovery encompasses two main strategies: the post-disaster reconstruction and post-disaster rehabilitation. Although they are the integral part of the post-disaster recovery process, they have different meanings. Post-disaster reconstruction means "the activity to revitalize and rebuild the disaster-stricken communities for a better living environment than they were before".[10] Post-disaster rehabilitation means "the disaster response activity that restores the basic services and facilities for the functioning of a community or a society affected by the disaster".[11]

The use of the urbanscape in the everyday life during the pre-disaster phase is related to the urban resilience during the disaster impact and the urban rehabilitation in the post-disaster recovery. The disability of a community to handle the consequences of the disaster impact derives from their vulnerable pre-disaster impact condition that allows them to be more exposed to the disaster impact. This vulnerable condition is influenced by many factors such as improper land use, the construction of informal settlements with self-built water supplies, environmental degradation, rapid and unplanned urbanization, poverty and climate change.[12] Due to the unplanned and rapid urbanization that influenced the construction of poorly planned housing settlements, in the period from 2000 until 2010, more than 680,000 people died in earthquake disasters.[13] All these factors lead to the urbanscape alteration that allows the disaster to have a more serious impact such as the organization and design of structures within open space causing direct landslides; rapid and unplanned urbanization, leading to high population density; and expansion of the urban fabric in the form of informal constructions on slopes or landfills.[14]

Every urbanscape has its unique characteristics and values. We perceive them with what we know and what we remember. We experience this through all of our senses and then interpret it through our instincts and intuitions.[15] This collective imagination is based on the urbanscape identity as our cultural inheritance that allows us to ascribe the urbanscape its values. The identity is shaped by the heritage as the richest historical record we possess, culture as the shaping force and the memory collected in these public spaces. Without the identity as a sense for belonging, we do not lose just a place, we simply loose ourselves. The identity depends on the quality of the open public places that raises the awareness to the perception and gives a sense of belonging.

[9] UNDP (2011:10)

[10] Yanase (2018:1)

[11] UNISDR (2017:7)

[12] Wisner and Adams (2002:13)

[13] Leoni et al. (2011:20)

[14] Wamsler and Brink (2015:17)

[15] Taylor (2008:1)

Since open public spaces as spaces of public values are exposed to the disaster effects, they undergo different forms of alteration that transforms the urbanscape and disrupts its identity. The word identity comes from the Latin word *identitas* and according to the Oxford English Dictionary it is described as "the fact of being who or what a person or thing is". The concept of identity has been used to describe the uniqueness of a person, by referring to the personal identity, social and place identity.[16] Hence to its broad usage, the concept of identity within this research refers to its spatial perception and interpretation within the urbanscape. The identity of place as the spatial perception of the identity concept is related to the "persistent sameness and unity that allows the place to be differentiated from other places", through the physical characteristics, accumulation of events and the people's meaning formed on the basis of their experiences.[17]

Therefore, the urbanscape identity is shaped by the heritage as the richest historical record we possess, culture as the shaping force and the memory that is collected in the open public spaces. Without the identity as a sense for belonging, we do not lose just a place, we simply lose ourselves. Relph describes this condition as *placelessness,* "the casual eradication of distinctive places and the making of standardized landscapes that results from an insensitivity to the significance of place".[18] The process of the urbanscape identity formation is strongly affected by the interaction between the people and their urbanscape. This interaction is reflected through the space identity that the people derive from their urbanscape, where they developed the sense for belonging and the spatial identity, when the people ascribe the identity to the urbanscape.[19] Both of these concepts are based on the two main identity classifications: the personal identity as the person's individual engagement with the surrounding and the collective identity, as the collective perception of the elements significant to the surrounding.[20] Followed with the disaster process, the disruption of the urbanscape identity changes the physicality of the urbanscape, as well as its perception and interpretation.[21]

Under the impact of natural forces, the urbanscape can be recovered and rebuilt in the post-disaster context, but affected by wars as intentional man-made forces, the urbanscape identity needs to be re-redefined, given a different meaning and reassessed in the post-disaster rehabilitation process. The deliberate destruction of cultural heritage during the war with the intention to erase the history and memory leaves an immense impact on the urbanscape identity. Therefore, the rehabilitation of war-torn urbanscapes requires a re-redefinition of its identity and to find the possibility to reconcile with the past. To redefine something means to change the meaning of something or to make people think about something in a new or different

[16] Kaymaz (2016:741)

[17] Seamon and Sowers (2008:45)

[18] Seamon and Sowers (2008:46)

[19] Stobbelaar and Pedroli (2011:323)

[20] Butler et al. (2017:12)

[21] Butler et al. (2017:5)

way.[22] This means that re-redefinition will be influenced by the perception and interpretation of the transformed urbanscape, its values and characteristics ascribed by the open public spaces. This network of open public spaces within the urban context contributes to the quality of the everyday life and at the same time plays an important role in the disaster management. Everyday gathering places were transformed into places for temporary shelters and evacuation routes. The people's perception and use of these spaces in everyday activities emphasize their inherited qualities in order to cope with the disaster situations.[23] During the war, open public spaces perform simultaneously the main target points of impact and the urban resilience line of protection. Their role of security and vulnerability transforms the urbanscape and the people's everyday life during the disaster. The everyday life pattern plays an important role in both disaster categories, the decision on evacuation routes in natural disaster and forms of resilience in man-made disasters. Therefore, by understanding their qualities in the everyday urban pattern in the pre-disaster phase, open public spaces can reduce the impact of the disaster during the disaster event phase and influence the post-disaster rehabilitation during the post-disaster phase.

1.1.1 The Main Research Problem and Research Question

The research discusses the possibility of recognizing the urban rehabilitation models of post-disaster urbanscapes, based on the comparative analysis of the chosen disaster case studies in the world, through the use of the urbanscape during the disaster process. The scope of this research is primarily concerned with understanding of how the urbanscape is used during the disaster process while addressing the main questions: How is the urbanscape network important in every phase of the disaster process? How each of the urbanscape elements embodies a different role during the disaster process? We seek to answer these questions with the aim to reveal the importance of the urbanscape, during the disaster process, based on its use and the interconnectedness of its elements between the disaster phases.

Based on the comparative analysis of the natural and man-made disaster process case studies, the research will define the criteria for future urban planning, based on the existing criteria for urban planning, derived from the urban rehabilitation models of urbanscapes exposed to disasters.

---

[22] https://dictionary.cambridge.org/dictionary/english/redefine

[23] Allan and Bryant (2010)

1.1.2 Research Method

The research will be conducted using the Heritage Urbanism (HERU) method[24] to reveal the urban rehabilitation models of urbanscapes exposed to disasters based on the comparative analysis of the significant post-disaster case studies in the world, both natural and man-made, in terms of the use of the urbanscape during the disaster process in each of the selected disaster categories in order to reveal the importance of the urbanscape during the disaster process, based on the use and the interconnectedness of its elements between the disaster phases.

The term Heritage Urbanism was created and developed as part of the research project titled Urban and Spatial Models for the Revival and Enhancement of the Cultural Heritage at the Faculty of Architecture of the University of Zagreb. The Heritage Urbanism method comprised of two main key words, cultural heritage and urbanism, with the aim to enhance cultural heritage, in search of revitalization and enhancement models. In order to recognize the models of the future use of heritage and interpret them as general models that can be applied to various specific cases, the Heritage Urbanism method is based on three methodological levels, "in terms of recognizing/determining/defining: the identity factors, factors of effect and value factors; evaluation criteria, enhancement criteria and the criteria for new interventions; and cultural heritage revival and enhancement models" (Obad Šćitaroci et al. 2019; Obad Šćitaroci and Bojanić Obad Šćitaroci 2019).

References

Aga E (2010) Networks of public open spaces in the urban fabric world 142:59–68. https://doi.org/10.2495/SW100061

Allan P, Bryant M (2010) The critical role of open space in earthquake recovery: a case study. NZSEE, Wellington

Antrop M (2005) Why landscapes of the past are important for the future. Landsc Urban Plan 70: 21–34. https://doi.org/10.1016/j.landurbplan.2003.10.002

Butler A, Sarlöv-Herlin I, Knez I, Ångman E, Ode Sang A, Åkerskog A (2017) Landscape identity, before and after a forest fire. Landsc Res 43:878–889. https://doi.org/10.1080/01426397.2017.1344205

https://dictionary.cambridge.org/dictionary/english/redefine. Accessed 30 Jun 2022

Kaymaz I (2016) Urban landscapes and identity in advances in landscape architecture. IntechOpen, London. https://doi.org/10.5772/51738

Leoni B, Radford T, Schulman M (2011) Disasters through a different lens. International strategy for disaster risk reduction. UNISDR, Geneva

Ministry of Home Affairs (2009) Understanding disasters. The National Institute of Disaster Management (NIDM), Delhi

Obad Šćitaroci M, Bojanić Obad Šćitaroci B (2019) Heritage urbanism. Sustainability 11:1–10. https://doi.org/10.3390/su11092669

[24] Heritage urbanism considers the revitalization and enhancement of cultural heritage in spatial, urban, and landscape contexts, and it explores models for its inclusion in contemporary life.

Obad Šćitaroci M, Bojanić Obad Šćitaroci B, Mrđa A (eds) (2019) Cultural urban heritage: development, learning and landscape strategies. Springer Nature, Cham. https://doi.org/10.1007/978-3-030-10612-6

Omićević N, Obad Šćitaroci BB (2018) The urbanscape duality during a disaster and post-disaster. Prostor 26:106–115. https://doi.org/10.31522/p.26.1(55).8

Saxena M (2016) Role of open spaces in disaster management. In: International conference AGORA 2016. G D Goenka University

Seamon D, Sowers J (2008) Place and placelessness, Edward Ralph. In: Key texts in human geography. Sage Publication, New York. https://doi.org/10.4135/9781446213742.n6

Stobbelaar D, Pedroli B (2011) Perspectives on landscape identity: a conceptual challenge. Landsc Res 36:321–339. https://doi.org/10.1080/01426397.2011.564860

Taylor K (2008) Landscape and memory: cultural landscapes, intangible values and some thoughts on Asia. In: Paper presented at the 16th ICOMOS General Assembly and International Symposium: finding the spirit of place—between the tangible and the intangible. ICOMOS, Quebec

UNDP (2011) Methodological guide for post-disaster recovery planning process. United Nations Development Program, New York

UNISDR (2009) UNISDR terminology on disaster risk reduction. United Nations international strategy for disaster risk reduction. UNISDR, Geneva

UNISDR (2017) Build back better in recovery, rehabilitation and reconstruction. Consultative version. UNISDR, Geneva

Wamsler C, Brink E (2015) The urban domino effect: a conceptualization of cities' interconnectedness of risk. In: Global assessment report of disaster risk reduction. International strategy for disaster risk reduction. UNISDR, Geneva

Wisner B, Adams J (2002) The nature of emergencies and disasters in environmental health in emergencies and disasters: a practical guide. World Health Organization, Geneva

Yanase N (2018) Land readjustment and post-disaster reconstruction in Japan, Chap. 2. In: Land readjustment: solving urban problems through innovative approach, 1st edn. Japan International Cooperation Agency Research Institute, Tokyo

Chapter 2
Literature

Abstract This chapter presents an overview of the most important findings of the analysed theoretical work, from different viewpoints. In the first part of the chapter, the literature related to the use of open public spaces in disaster management is discussed, as the starting point of the research. The second part of the chapter describes the literature related to the urbanscape identity before and after the disasters. The last, third part of the chapter outlines definitions and concepts in disaster research, based on the interdisciplinary approach.

Keywords Disasters · Open public spaces · Disaster management · Urbanscape identity

2.1 Literature

2.1.1 The Use of Open Public Spaces in the Disaster Management

Extensive research has shown how the network of open public spaces, as the basic structural elements of the urban fabric (Aga 2010), reflects the character of the city (Saxena 2016). Studies of the network of open public spaces show how the clarity of their structure and accessibility plays an important role in the formation of the urbanscape and influences the ability to reinforce the relationship between those spaces and the people who use them (Aga 2010).

In terms of examining the use of open public spaces during the disaster process, most studies have been focused only on the role of the urbanscape in the disaster management during the disaster event phase. Their use during the entire disaster process that embodies the role of duality (Omićević and Bojanić Obad Šćitaroci 2018) has not been given sufficient consideration in terms of vulnerability in the pre-disaster phase that can reduce or increase the disaster impact; enhancing the role of security and vulnerability during the disaster impact; and enhancing the role of rehabilitation and the opposing indirect effects in the post-disaster recovery process. Most of the existing research in the context of the post-disaster recovery process is

N. Omićević, B. Bojanić Obad Šćitaroci, *The Urban Rehabilitation of Post-Disaster Scapes*, Disaster Resilience and Green Growth,
https://doi.org/10.1007/978-981-19-9505-7_2

mainly focused on the analysis of the numbers related to the impact on population and property along with estimated economic costs.

A few published studies have examined the role of open public spaces and their possibilities in terms of handling the disaster impact, which show how the open spaces have contributed to lower the disaster risks (Saxena 2016). This is visible in their role in prevention of floods, through the maintenance of wetlands; in their role in rescue and recovery operations during floods, through the large tracts of open land and urban infrastructure that serve as open pockets for the evacuation; in the need of having open spaces within the surrounding environment of buildings in case of an earthquake; and in the need of having open spaces in the forest to prevent the fire spreading in the cases when fire breaks and to reduce the impact of future wildfires (Saxena 2016). The use of the urbanscape and its network contributes to the quality of the everyday life and at the same time plays an important role in the disaster process (Saxena 2016; Omićević and Bojanić Obad Šćitaroci 2018). The normal urban pattern of the everyday life has been analysed to show its importance during the disaster impact in form of urban resilience to cope with the disaster effects (Lacan and McBride 2009), but the question of how these open public spaces are being differently altered, when influenced by the natural or man-made disaster impact, has not been addressed in previous studies.

2.1.2 The Urbanscape Identity Before and After the Disasters

The existing body of research on the relation between the place identity, spatial identity and the urbanscape shows the concept of identity by revealing how the meaning and experience play an important role in the perception of the identity (Butler et al. 2017; Kaymaz 2016) and the landscape identity before and after the disaster, by revealing how the physicality of the landscape changes in the aftermath of the disaster together with its perception and experience, leading to the disruption of its identity (Wisner and Adams 2002). Results from earlier studies demonstrate the possible changes that can impact the urbanscape, but they have failed to find a link between these changes that lead to the transformation of the urbanscape identity and its redefinition.

2.1.3 Definitions and Concepts in Disaster Terminology

The term disasters is used by diverse scientific disciplines and includes varied meanings. For the purpose of the research, disasters are being interpreted in order to understand its diverse concept and definition. The contemporary interpretation was based on the interdisciplinary approach, through the scope of the following disciplines: Social and Cultural, Historical, Environmental, Landscape, Urbanism and Economy.

The term disaster is used to describe an event that results in harm, damage or death. From the social aspect, disasters occur with the disability of a community to cope with an extreme event. This disability is connected with the pre-disaster impact condition that allows the society to become vulnerable. From the perspective of social vulnerability, this disability refers to the "people's capacity to anticipate, cope with, resist and recover from the disaster impact" (Federal Emergency Management Agency FEMA 2014). The economic aspect is visible through the property damage, financing recovery costs for the damage assessments, debris removal, infrastructure restoration and the re-planning of the disaster stricken areas (Federal Emergency Management Agency FEMA 2014). Improper land use and deforestation are some of the factors that increase the vulnerability to the disaster impact. These factors influence the environmental aspect of disasters and refer to the inadequate drainage of self-built water supplies, poor maintenance of septic tanks that lead to the instability of the hills, rapid unplanned urbanisation and climate change (Federal Emergency Management Agency FEMA 2014). Since the character of the city is reflected through the buildings and their forms and the open spaces, the urban aspect of disasters is reflected through the attributes of these open spaces, in terms of handling the disaster impact (Saxena 2016). The open spaces consist of the organised open and natural landscape that can lower the disaster risks in form of prevention to reduce the disaster impact. They play an important role in the rescue operation, resilience and recovery process (Saxena 2016; Pilav 2012) and shape the patterns of disaster commemorations, inscribed within the landscape. "Disasters simultaneously re-shape the physical and symbolic landscapes", through memorials, that interpret and contextualize the disaster experience (Wamsler and Brink 2015). In terms of the landscape aspect, decreasing the disaster impact is possible through the notion of mitigation, such as mitigation of floods, through wetlands, enabling the continuity of fuels, in forms of controlled burning of vegetation within a forest, to prevent wild fire (Saxena 2016). The landscape functions as a method of transformation in the disaster context, which is influenced by the way people perceive and use the landscape attributes (Villagra-Islas and Alves 2016). From the historical aspect, the cultural heritage enhances the disaster resilience to reduce the disaster impact (Murthy 2013). Resilience is defined as "the ability of a system and its components to anticipate, absorb and accommodate or recover from the effect of a hazardous event in a timely and efficient manner, through ensuring preservation, restoration and improvement of its essential basic structures and functions" (Murthy 2013). Disasters caused by intentional forces, such as war, aim to impact the heritage sites in order to erase the evidence of history. The loss of cultural properties and their value leads to the loss of identity as a sense of place. A well-conserved historic environment reduces the factors of vulnerability and strengthens the resilience of the society to the disaster impact (Murthy 2013).

References

Aga E (2010) Networks of public open spaces in the urban fabric world. Sustain World 142:59–68. https://doi.org/10.2495/SW100061

Butler A, Sarlöv-Herlin I, Knez I, Ångman E, Ode Sang A, Åkerskog A (2017) Landscape identity, before and after a forest fire. Landsc Res 43:878–889. https://doi.org/10.1080/01426397.2017.1344205

Federal Emergency Management Agency FEMA (2014) Hazard, vulnerability and risk analysis. In: Chap. 6: Social impact of disaster. Federal Emergency Management Agency FEMA, Washington, DC, pp 153–191

Kaymaz I (2016) Urban landscapes and identity in advances in landscape architecture. IntechOpen, London. https://doi.org/10.5772/51738

Lacan I, McBride J (2009) War and trees: the destruction and replanting of the urban and peri-urban forest of Sarajevo, Bosnia and Herzegovina. Urban For Urban Green 8:133–148. https://doi.org/10.1016/j.ufug.2009.04.001

Murthy M (2013) Heritage and resilience. Issues and opportunities for reducing disaster risks. International strategy for disaster risk reduction. UNISDR, Geneva

Omićević N, Bojanić Obad Šćitaroci B (2018) The urbanscape duality during a disaster and post-disaster. Prostor 26:106–115. https://doi.org/10.31522/p.26.1(55).8

Pilav A (2012) Before the war, war, after the war: urban imageries for urban resilience. Int J Disaster Risk Sci 3:23–27. https://doi.org/10.1007/s13753-012-0004-4

Saxena M (2016) Role of open spaces in disaster management. AGORA, Baltimore

Villagra-Islas P, Alves S (2016) Open space and their attributes, uses and restorative qualities in an earthquake scenario: the case of concepcion. Chile 19:56–67. https://doi.org/10.1016/j.ufug.2016.06.017

Wamsler C, Brink E (2015) The urban domino effect: a conceptualization of cities' interconnectedness of risk. In: Global assessment report of disaster risk reduction. International strategy for disaster risk reduction. UNISDR, Geneva

Wisner B, Adams J (2002) The nature of emergencies and disasters in environmental health in emergencies and disasters: a practical guide. World Health Organization, Geneva

Chapter 3
Research on the Disaster Process Case Studies in the World

Abstract This chapter presents an analytical overview of the disaster process case studies in the world, based on the criteria for the selection of the natural and man-made disaster case studies in the world and the criteria for the analysis of the disaster process case studies in the world.

Based on the criteria for the selection of the natural and man-made disaster case studies, the research is based on the comparative analysis of the natural disasters, which include four main disaster types, analysed through their sub-categories: hydrological, meteorological, climatological and geophysical disasters. For the purpose of the research, the comparative analysis of the man-made disasters includes the sociological and technological disaster type, analysed through its sub-categories.

Based on the criteria for the analysis of the disaster case studies in the world, the research includes the analysis of 18 disaster case studies in the world, of which 12 are case studies within the natural and six case studies within the man-made disaster category, based on the natural and man-made disaster subcategory and the period of their occurrences from 1991 to 2022.

Keywords Disaster process · Natural disaster · Man-made disasters

3.1 Research on the Disaster Process Case Studies in the World

Disasters are not described through one physical event; they evolve as a combination of three factors: the event of hazards confronted with the degree of vulnerability and the capacity to mitigate the negative consequences of the disaster risk that resulted from the interaction of the hazards and the vulnerability. The main impact factors that can contribute to increasing the disaster risk include: the climate change that causes new hazards; poverty and socio-economic inequality; rapid urbanization and the inadequate access to urban infrastructure and facilities. By the year 2030, it is expected that five billion of the planet's population of 8.1 billion will live in an urban environment, causing environmental degradation, by devastating the natural

N. Omićević, B. Bojanić Obad Šćitaroci, *The Urban Rehabilitation of Post-Disaster Scapes*, Disaster Resilience and Green Growth,
https://doi.org/10.1007/978-981-19-9505-7_3

environment such as the forest and wetlands that can contribute to protect and decrease the disaster risk.[1]

For the purpose of obtaining the credibility of the conclusions that will derive from the comparative analysis, the research will include the analysis of 18 disaster case studies in the world, of which 12 case studies are within the natural and six case studies within the man-made disaster category. The criteria for choosing these examples will be based on the division into natural and man-made disaster sub-category and the period of their occurrence, from 1991 to 2022. The main purpose of this comparative analysis is to reveal the use of the urbanscape during the disaster process, through the role of its open public spaces during each disaster phase, in both, natural and man-made disaster categories.

3.1.1 Criteria for the Selection of the Natural Disaster Case Studies in the World

Natural disasters can be sub-divided into four main disaster types: hydrological disasters, that include floods, landslides and mudflows and avalanches; meteorological disasters, that include storms, typhoons, cyclones, hurricanes, winter storms, tornadoes and tropical storms; geophysical disasters, that include earthquakes, volcanic eruptions and tsunamis; and climatological disasters, that include droughts, extreme temperatures and wildfires.[2,3]

Hydrological Disasters

Hydrological disasters, with 164 floods and 13 landslides, have been recorded as the most common within the natural disaster occurrences in 2016.[4] Floods can be induced by heavy or continuous rain that exceeds the assortative capacity of the soil or the river flow capacity, streams and coastal areas. They can arise as a result of population growth and urbanization, environmental degradation through the loss of forests and natural vegetation and climate change. The measures that can reduce the risk of flooding involve the development of new building codes that could reinforce the flood resistance, avoiding constructions on flood-prone land and creating more spaces for rivers and wetlands that could contribute to the flood mitigation by reducing the speed of seawater, strength of the waves and the force of the wind.[5]

---

[1] Brigitte et al. (2011:33)
[2] Guha-Sapir et al. (2004:22)
[3] Guha-Sapir et al. (2017)
[4] Guha-Sapir et al. (2017:33)
[5] Brigitte et al. (2011:137)

Landslides, just like floods, can be induced by natural forces: heavy rain, snow melts, earthquakes and volcanic eruptions, but they can also be triggered by human act, by deforestation and constructions on unstable slopes. Deforestation and the inappropriate land use that leads to the instability of the slopes, followed by the high population density increases the future risk of landslide occurrence. Measures that can reduce the risk of landslides include construction of concrete retaining walls to support the slopes, improving the drainage to stabilize the slopes, reinforcing the riverbanks, protecting the forest coverage and using building materials that could contribute to the infrastructure resilience to the disaster risk.[6] In both disaster sub-categories, urban areas are the most vulnerable to future disaster risk, mainly because of the population density and the people that were forced to live on unstable lands.

Within the hydrological disaster natural subcategory, the following disaster case studies will be analysed: Floods along the Gulf Coast, the United States of America, (2005); Floods and Landslides in Switzerland, (2005); and Floods in Pakistan, (2010).

Meteorological Disasters

Meteorological disasters, out of 83 storms, had a share of 28.1% of the total natural disaster occurrences in 2016.[7] Although they are named differently, hurricanes, cyclones and typhoons are considered to be the same disaster type, referring to the "cyclones in the Indian Ocean and the South Pacific; hurricanes in the western Atlantic and eastern Pacific; and typhoons in the western Pacific".[8] Hurricanes are considered to be more destructive than floods since the property and population damage encompasses not only the coast line but the inland as well. The fragile settlements and the population living along the coastal area are exposed the most to extreme wind speeds and therefore are the most vulnerable to the future disaster risk. Measures that need to be undertaken in order to reduce the impact of storms include: using flood resistance building materials; integrating the flood risk assessment into the urban planning strategies; avoiding construction of buildings directly along the coast line; protecting trees and coral reefs, since they can contribute to the disaster mitigation in form of natural wave breakers; and preservation of wetlands and forests that can store large amounts of floodwater and contribute to the disaster mitigation as well.[9]

The following disaster case studies will be analysed within the meteorological disaster natural subcategory: Cyclone in Bangladesh, (1991); Cyclone in Australia, (2017); and Hurricane in the United States of America, (2012).

---

[6] Brigitte et al. (2011:144)

[7] Guha-Sapir et al. (2017:34)

[8] Brigitte et al. (2011:140)

[9] Brigitte et al. (2011:142)

Geophysical Disasters

Geophysical disasters had an overall share of 9.1% of the total natural disaster occurrences in 2016. Out of the 31 registered geophysical disasters, 30 were caused by earthquakes.[10] The Earth surface consists of 15 tectonic plates, the movement of which may induce an earthquake. About 90% of earthquake occurrences are of tectonic origin.[11] Although the places of their occurrences can be identified, it is difficult to predict the time or scale of their impact. The most vulnerable areas are the urban centres and the population living in earthquake prone areas. Due to their economic constraints, people are forced to live in crowded and dense areas and construct housing that is unable to resist the seismic shocks, which leads them to be more exposed to the disaster risk. Measures that can reduce the disaster risk involve integration of seismic risk into the urban land-use planning, especially in urban centres, to avoid dense areas and leave sufficient open space between the buildings for evacuation; enforcement of the building infrastructure and their use in the aftermath of the earthquake; and the reduction of fire risk in the disaster aftermath.[12]

Volcanoes are considered ruptures in the Earth surface, which produce the eruption of the ash and gases from the hot molten rock below the surface. There are more than 1500 potentially active volcanoes, where 50–60 volcanoes erupt every year worldwide.[13] The most vulnerable areas exposed to the volcano risk are the settlements on the volcano slopes and valleys. More than 60 large cities in the world live near potentially active volcanoes.

Measures that need to be undertaken to reduce the risk of volcano eruptions include the integration of volcano risk in the land-use planning to avoid constructions and the development of infrastructure in volcano hazardous areas; to ensure evacuation and response plans in the urban strategies; and to install a monitoring system that could predict possible eruptions.

Within the geophysical disaster natural subcategory, the following post-disaster case studies will be analysed: Volcanic eruption of Mt. Pinatubo in the Philippines (1991); Earthquake and Tsunami in Japan (2011) and Earthquake and Landslides in Nepal (2015).

Climatological Disasters

Climatological disasters had a share of 11.1% of all natural disaster occurrences in 2016 and reported 38 occurrences in total.[14] Wildfires as climatological disasters may not have the same population impact as the other natural disasters, but in terms

---

[10] Guha-Sapir et al. (2017:37)

[11] Brigitte et al. (2011:133)

[12] Brigitte et al. (2011:134)

[13] Brigitte et al. (2011:154)

[14] Guha-Sapir et al. (2017:36)

of the property impact, they can be very destructive. Wildfires are described as uncontrolled fires that have a devastating impact on the forest and the vegetation and animal species within. They can be induced by fuel, oxygen and heat source and cause land degradation that can activate more floods and landslides.[15] Measures that need to be undertaken to reduce uncontrolled fires include building of fire breaks within the forest, if no natural fire break exist; clearing the vegetation in the surrounding settlements within a fire risk hazardous area; planting vegetation of low flammability and using fire-resistant building materials and advanced methods in the prescribed burning control.[16]

Within the climatological disaster natural subcategory, the following post-disaster case studies will be analysed: Wildfires in Indonesia (1997); Wildfires in California, the United States of America (2017, 2018) and Heat Wave and Wildfires in Russia (2010).

3.1.2 Criteria for the Selection of the Man-Made Disaster Case Studies in the World

Man-made disasters have been defined as "emergency situations of which principal, direct causes are identifiable human actions, deliberate or otherwise".[17] Man-made disasters can be caused by intentional human actions, such as terrorism, conflicts of war and genocide and non-intentional human actions, such as technological or industrial disasters. For this reason, they can be sub-divided into two main disaster types: sociological and technological man-made disasters.[18] Technological disasters occur as the result of failure of technology, while the sociological disasters with their intentional nature are described as an urban act of terror.[19]

For the purpose of this research, the comparative analysis of the man-made disaster case studies will be focused on the sociological man-made disaster, referring to "the urban wars of the twenty-first century"[20] and the technological disasters, referring to industrial disasters, transport accidents and chemical, nuclear or biological disasters.[21]

---

[15] Brigitte et al. (2011:158)

[16] Brigitte et al. (2011:159)

[17] Al-Dahash et al. (2015:435)

[18] Al-Dahash et al. (2015:436)

[19] Beall (2007)

[20] Beall (2007:4)

[21] Loenhout et al. (2020:2)

Sociological Man-Made Disasters

War is defined as an "organized and often prolonged conflict, characterized by extreme violence, social disruption and economic destruction", while terrorism is defined as the "threatened use of force or violence against people to achieve political, religious or ideological objectives".[22] During the impact of sociological disasters, cities have always been sites of protection and sites of attack, because the struggles for survival and power occur within the urban space that enhances the spatial dimension of the social interaction within and their capability to confront the reality.[23] During the disaster impact, urban spaces become urban targets or areas of terror, but they also become sites of reconstruction and peace-building. For this reason, the research will be focused on analysing the importance of the urban dimension of the conflicts.

Within the sociological man-made disaster subcategory, the following disaster case studies will be analysed: The Siege of Sarajevo (1992–1996); The Battle of Grozny (1994–1995) and The September 11 Terrorist Attacks in New York (2001).

Technological Man-Made Disasters

Technological disasters are defined as "those induced entirely or predominantly by human activity and choices".[24] Technological disasters arise from technological or industrial conditions, such as dangerous procedures, infrastructure failures or specific human activities that can lead to population loss, property damage, economic disruption or environmental risk.[25] Further sub-divided, industrial disasters refer to chemical spills, explosions, fire, gas leaks, poisoning radiation and other, while transport disasters include accidents in air, on the rail, road and water.

Technological disasters encompass 36.4% of all reported disasters worldwide.[26] According to the EM-DAT database,[27] from 2000 until 2019, 5143 technological disasters were recorded, out of where 3532 were reported as transport disasters, which affected more than 1.4 million people.[28] Due to their random and unpredictable nature of occurrence, their effect on the population can be long lasting and stressful. The rapid and poorly planned growth of the urban areas with intense industrial activity within increases the impact of technological disasters. Therefore,

---

[22] Al-Dahash et al. (2015:436)

[23] Beall (2007:4)

[24] UNISDR (2018:9)

[25] UNISDR (2009:17)

[26] Skilodimou and Bathrellos (2021:1)

[27] The EM-DAT database covers natural and technological disasters, occurrences and effects, from 1900 to present (ongoing), maintained by the Centre for Research on the Epidemiology of Disasters (CRED).

[28] Loenhout et al. (2020:2)

appropriate urban planning is the key element to reduce the impact of the technological disaster events. For this reason, the research will be focused on analysing the impact of the technological disasters and its effects on the urban environment.

Within the technological man-made disaster sub-category, the following disaster case studies will be analysed: Deepwater Horizon Oil Spil in the Gulf of Mexico (2010); Covid-19 Pandemic (2019-Ongoing) and Beirut Port Explosion (2020).

3.1.3 Criteria for the Analysis of the Disaster Process Case Studies

The criteria for re-redefining the urbanscape identity will be carried out based on the comparison of the selected post-disaster case studies, through the analytical overview of the disaster process, based on the use of the urbanscape, through its built and natural environment, reflected through the following urbanscape elements: the urban pattern of the everyday life and the urban vulnerability during the pre-disaster phase; the urban impact and the urban resilience during the disaster event phase; and the urban rehabilitation strategies during the post-disaster phase.

As stated in the introduction, the urbanscape in this research is defined as the open public spaces within the city that include the natural and the built environment of the city. Therefore, for the purpose of this research, the urbanscape built environment will be analysed through its open public places and the urban fabric components (physical properties, urban infrastructure) while the natural environment will be analysed through its green spaces (parks, greenways, rivers, hillsides) and the natural landscape which provides the urbanscape sustainability.

The main aim of this comparative analysis is to reveal the importance of the urbanscape during each phase of the disaster process, by focusing on the role of each element that comprises its natural and built environment and to reveal the urbanscape identity disruption, derived from the disaster effects with the aim of showing how the classification of the natural and man-made disaster effects influences the rehabilitation of the disrupted urbanscape identity in the post-disaster context.

Therefore, the comparison of the selected post-disaster case studies will be based on the natural and man-made disaster sub-category division and the period of their occurrence, from 1991 to 2022, through the analytical overview of each urbanscape element that comprise its natural and built environment. Although the selected disaster case studies cover different scales of spatial extent, from a region to a city, which is being exposed to the disaster impact, the research defines and explores only the function of each urbanscape element in the pre-disaster phase, during the disaster event phase and the post-disaster phase, so these can be applied as separate measures and reveal the system of the existing criteria for urban planning in the natural and man-made disaster process case studies, regardless of the scale of the spatial extent being exposed to disasters.

The Pre-disaster Phase of the Disaster Process

The pre-disaster phase enhances the process of reduction as a protective and preventive action, required to reduce the disaster impact. This process consists of mitigation tools that need to be undertaken to reduce the disaster effects and decrease the vulnerability for future disaster risks. One of the most important tools that can reduce such damages, reflected through the use of the urbanscape, is the awareness. As a state of being aware, awareness can be achieved by identifying the quality of the open public spaces that form the urbanscape in the pre-disaster phase or the everyday life. The quality of these open public spaces affects their use that can increase the possibility to cope with the disaster impact and decrease their vulnerability. Vulnerability is described as "the circumstances of a system that make it susceptible to damaging effects of a hazard".[29] The disability of a community to handle the consequences of the disaster impact is derived from its vulnerable pre-disaster condition that increases the disaster impact. This vulnerable condition is influenced by many factors, such as physical, social, economic and environmental factors, but for the purpose of this research, their vulnerability will be analysed through the urbanscape elements. This vulnerability has been developed with regard to the urbanscape as urban vulnerability, such as improper land use, the construction of informal settlements with self-built water supplies, environmental degradation and rapid and unplanned urbanization.

Since the function and use of the open public spaces in the everyday life can contribute to prevent and reduce the disaster impact, the comparative analysis of the chosen disaster case studies in their pre-disaster phase of the disaster process will be conducted through the process of prevention and mitigation based on urban pattern of the everyday life and their factors of existing urban vulnerability that could reduce or increase the disaster impact.

The Disaster Event Phase of the Disaster Process

Being defined through their risk and vulnerability, disasters can lead to a short-term and long-term impact on the urbanscape and society, by causing direct, secondary and indirect effects. Direct effects are reflected through injuries and death and the physical damage and destruction. The secondary effects evolve from the initial disaster impact that leads to a combined disaster impact. Indirect effects derive from the direct and secondary effects and they are the result of the interrelation between the society and the urbanscape in terms of evacuation and displacement of population, urban sprawl and urban growth, informal constructions and environmental challenges, etc. For the purpose of the research, the disaster impact refers not only to the population and property impact, it is also related to the urban impact and urban resilience based on the impact that encompasses the urbanscape and its

[29] UNISDR (2009:30)

resilience. The concept of resilience has been developed as framework for the disaster risk management. The word resilience comes from the Latin word *resalire*, "the idea of resisting and rebounding".[30] With regard to the urbanscape, the concept of urban resilience is not related to the city's ability to return to its previous state, but to achieve a state that could adjust, regardless if it differs from the initial state. Urban resilience refers to the ability to develop in a reduced capacity, while maintaining the urbanscape main functions, even if some of its elements get disrupted.[31]

Therefore, the comparative analysis of the chosen disaster case studies in the disaster event phase will be conducted through the direct disaster effects, based on the population impact, property impact, urban impact and the urban resilience; the secondary disaster effects and the indirect disaster effects.

The Post-disaster Phase of the Disaster Process

Followed with the indirect disaster effect, the post-disaster phase is reflected through an ongoing recovery process, with the aim to use the change and turn it into sustainable and long-term development as the development that meets the needs of the present without compromising the ability of future generations to meet their own needs. As a link between the disaster response and the long-term sustainable development, the post-disaster recovery encompasses two main strategies: the post-disaster reconstruction and post-disaster rehabilitation. Although they are the integral part of the post-disaster recovery process, they have different meanings. Post-disaster reconstruction means the activity to revitalize and rebuild the disaster-stricken communities to what they were before. Post-disaster rehabilitation means the disaster response activity that restores the basic services and facilities for the functioning of a community or a society, affected by a disaster. The post-disaster recovery process of rehabilitation includes three main types: physical, social and psychological rehabilitation. For the purpose of this research, the physical rehabilitation refers to the urban rehabilitation as the reconstruction of the urban fabric components; the physical properties, the urban infrastructure and the natural landscape, all of which is reflected through the influence of the urbanscape in the post-disaster recovery process.

Therefore, the comparative analysis of the chosen disaster case studies in their post-disaster phase of the disaster process will be conducted through their post-disaster recovery process based on the rehabilitation plan, rehabilitation program and the urban rehabilitation strategies used. The post-disaster rehabilitation will be reflected through possible post-disaster urban rehabilitation models, resulted from the comparative analysis of the disaster case studies in the world.

The comparative analysis of the selected disaster case studies is summarized in Table (01–18) for each case study disaster category, in form of an overview of the

---

[30] Leroy et al. (2016:2)

[31] Leroy et al. (2016)

disaster process, which is based on the use of the urbanscape, through its built and natural elements: the urban pattern of the everyday life and the urban vulnerability during the pre-disaster phase; the direct urban impact during the disaster event phase; and the urban rehabilitation during the post-disaster phase.

References

Al-Dahash H, Kulatunga U, Al-Dehesh A (2015) Disaster response management stemming from war operations and terrorism in Iraq. In: Conference paper in the 12th international postgraduate research conference (IPGRC 15). MediaCityUK, Manchester

Beall J (2007) Cities, terrorism and urban wars of the twenty-first century. Crisis States Research Centre, London

Brigitte L, Radford T, Schulman M (2011) Disasters through a different lens. International strategy for disaster risk reduction. UNISDR, Geneva

Guha-Sapir D, Hargitt D, Hoyois P (2004) Thirty years of national disasters 1974–2003: the numbers. Centre for Research on the epidemiology of disasters (CRED). Presses universitaires de Louvain, Brussels

Guha-Sapir D, Hoyois P, Wallemacq P, Below R (2017) Annual Disaster Statistical Review 2016. Centre for research on the epidemiology of disasters (CRED). Universite catholique de Louvain, Brussels

Leroy J, Amdal J, Cariolet J, Diab Y, Becue V (2016) Vulnerability and social resilience: comparison of the neighbourhoods in New Orleans after Hurricane Katrina. In: Conference paper presented at the European Conference on Flood Risk Management. EDP Sciences, Paris. https://doi.org/10.1051/e3sconf/20160708002

Loenhout J, Below R, Horions M (2020) Technological disasters. Centre for research on the epidemiology of disasters (CRED). presses universitaires de Louvain, Brussels

Skilodimou H, Bathrellos G (2021) Natural and technological hazards in urban areas: assessment, planning and solutions. MDPI 13:1–5. https://doi.org/10.3390/su13158301

UNISDR (2009) UNISDR terminology on disaster risk reduction. United Nations international strategy for disaster risk reduction. UNISDR, Geneva

UNISDR (2018) Man-made and technological hazards. International strategy for disaster risk reduction. UNISDR, Geneva

Chapter 4
Comparative Analysis of the Selected Hydrological Disaster Process Case Studies

Abstract This chapter presents an analytical overview of the selected hydrological disaster case studies in the world that include floods, landslides, mudflows and avalanches. Within the hydrological disaster natural subcategory, the following disaster case studies are analysed: Floods along the Gulf Coast, the United States of America (2005); Floods and Landslides in Switzerland (2005) and Floods in Pakistan (2010), based on the use of the urbanscape, through its built and natural elements: the urban pattern of the everyday life and the urban vulnerability during the pre-disaster phase; the direct urban impact during the disaster event phase and the urban rehabilitation during the post-disaster phase.

Keywords Natural disasters · Hydrological disasters · Floods · Landslides

4.1 Comparative Analysis of the Selected Hydrological Disaster Process Case Studies

4.1.1 Floods Along the Gulf Coast, the United States of America (2005)

The Pre-disaster Phase of the Disaster Process

With the aim of building resilience to the impact of flood disasters, in 1959, "the Standard Project Hurricane" was developed. The project consisted of raising levels and building floodwalls that would protect the cities from a 200-year storm.[1] This protection system failed with the impact of hurricane Katrina. As one of the strongest storms in the past 100 years in the United States, on August 29, 2005, Hurricane Katrina hit the Gulf Coast of the United States.[2] The hurricane became an organized system from August 23, by reaching a Category 5 hurricane, until it made landfall on

[1] Leroy et al. (2016:5)

[2] Leroy et al. (2016)

N. Omićević, B. Bojanić Obad Šćitaroci, *The Urban Rehabilitation of Post-Disaster Scapes*, Disaster Resilience and Green Growth,
https://doi.org/10.1007/978-981-19-9505-7_4

August 29, as a Category 3 hurricane. The storm moved the next 2 days as land-falling hurricane and became downgraded on August 31.[3] Floods can be mitigated by maintaining the wetlands with vegetation, since they can store water. The vegetation along the waterside decreases the level of the rising flood, through the absorption by their root system. Therefore, the treatment of the open public spaces in the pre-disaster phase influences the flood resistance, avoidance and resilience.[4] One of the main pre-disaster factors that increased the flood level was the shallowness and the urban shape of the shoreline[5] and the factor that contributed to the vulnerability to the impact was the lack of the population risk awareness.[6]

The Disaster Event Phase of the Disaster Process

The land-falling hurricane produced 62 tornadoes in eight states and a high rainfall that resulted in destructive floods. This disaster caused the death of 1353 people and the damage to and destruction of 275,000 homes.[7] The waves of the storm caused severe damage to buildings in coastal areas. Due to the high wind and poor housing construction, the hurricane caused damage and destruction of building envelopes and rooftop equipment. The hurricane caused the modification of coastal areas, as the result of the flood, high waves, waterborne contamination, and severe erosion. The main communication facilities and vehicles were damaged or destroyed.[8] The city's flood protection system was based on raising flood walls and levees, slowing down water flows with canals and establishing pumping stations, the resilience of which failed. This protection system broke in over 50 locations. The levees and the floodwalls broke, resulting in flooding of 80% of the urban fabric. Although many people managed to be evacuated, more than 30,000 evacuees that stayed behind were without enough supplies for days.[9] They were forced to evacuate onto higher levels or to find rescue in shelter zones. The unbuilt open public spaces during the disaster impact enabled the possibility for establishing shelter zones with rescue tents as medical facilities.

The Post-disaster Phase of the Disaster Process

The post-disaster rehabilitation process consisted of three main rehabilitation programs: the establishment of the Bring New Orleans Back Commission comprising of

[3] NOAA (2006:5)

[4] Saxena (2016:4)

[5] Omićević and Bojanić Obad Šćitaroci (2018)

[6] Leroy et al. (2016)

[7] NOAA (2006:1)

[8] FEMA (2006)

[9] Leroy et al. (2016:5)

professional planners and designers with the main rehabilitation strategy to develop a citywide plan by proposing urban design and land-use solutions that would reduce the risk of future flooding. This strategy included shrinking the city's footprint and replacing the neighbourhoods from the lower land level with green spaces; the redevelopment of the neighbourhood plans for the flooded areas within "The Neighbourhood Rebuilding Plan" of the City Council and "The Unified New Orleans Plan", that consisted of three phases. These phases were developed on two spatial scales, the district and the citywide scale, and included: the recovery assessment, the recovery scenario and proposing the strategic recovery framework.[10] In June 2006, the "Hurricane Katrina Memorial Commission"[11] was developed, with the main aim to create a permanent memorial dedicated to the events and deaths of the disasters as the main tool of resistance in the rebuilding process of the city of New Orleans and a park and mausoleum for the unidentified bodies, found in the aftermath of the storm.[12] Despite these formal memorials, the post-disaster process was also marked with many informal monuments and artworks of different styles and materialization.

The Post-disaster Urban Rehabilitation Models

The disaster process overview of the floods along the Gulf Coast reveals several post-disaster urban rehabilitation models. Although the protection system of the Standard Project Hurricane from 1959 failed to resist the impact of Hurricane Katrina, the strategic urban tools in raising levels and building floodwalls to build resilience to future floods reveal the post-disaster urban rehabilitation model of urban resilience. Another strategic tools that were used to reduce the risk of future floods referred to reshrinking the city's footprint and replacing the neighbourhoods in low land with green spaces and vegetation with the ability to store water, revealing the post-disaster urban rehabilitation model of shrinking the city's footprint and land-use relocation. These post-disaster urban rehabilitation strategies implied the redevelopment of the neighbourhood plans in the flooded areas, revealing the post-disaster urban rehabilitation model of urban plan redevelopment. With the appearance of many formal and informal memorials in the aftermath of the disaster, the post-disaster rebuilding process of New Orleans is reflected through the post-disaster urban rehabilitation model of memorialization using monuments and memorials (Table 4.1; Fig. 4.1).

[10] Kondo (2008:5)

[11] Morris (2009:157)

[12] Morris (2009:158)

Table 4.1 Disaster process overview of the floods along the Gulf Coast, the United States of America

Disaster process overview			
Disaster category		Natural	
Disaster subcategory		Hydrological disasters: floods	
Case study disaster (country, region/city, date, year of disaster event)		United States of America, Gulf coast, August 23–31, 2005	
Disaster process overview—use of the urbanscape			
The pre-disaster phase Prevention and mitigation			
Urban pattern of the everyday life		Urban vulnerability	
Wetlands can store water and with their maintenance through the vegetation along the waterside, floods can be mitigated The treatment of the open public spaces in the pre-disaster phase influences the flood resistance, avoidance and resilience		The existing shallowness and urban shape of the shoreline increased the flood level The lack of the population risk awareness Existing protection system that failed	
The disaster event phase Disaster impact—urbanscape identity disruption			
Direct disaster effect		Secondary effect	Indirect disaster effect
Population impact	1353 people died	Landscape alteration in terms of the modification of the coastal area Disabling the function of urban infrastructure	Relocation of population
Property impact	The damage and destruction of 275,000 homes The waves of the storm caused severe damage to buildings in coastal areas. Due to the high winds and poor housing construction, the hurricane caused damage and destruction to building envelopes and rooftop equipment		
Urban impact	The hurricane caused modification of coastal areas, as the result of the flood, high waves, waterborne contamination, and severe erosion The main communication facilities and vehicles were damaged or completely destroyed The existing protection system broke in over 50 locations. The levees and the floodwalls broke, resulting in flooding of 80% of the urban fabric		
Urban resilience	The city's flood protection system was based on raising flood walls and levees, slowing down water flows with canals and establishing pumping stations, the resilience of which failed Evacuation of the evacuees that		

(continued)

Table 4.1 (continued)

The disaster event phase Disaster impact—urbanscape identity disruption			
Direct disaster effect		Secondary effect	Indirect disaster effect
	stayed was based on finding rescue on higher levels or in shelter zones. The unbuilt open public spaces during the disaster impact enabled the possibility for establishing shelter zones with rescue tents as medical facilities		

The post-disaster phase Post-disaster recovery—post-disaster urbanscape identity rehabilitation		
Rehabilitation program	Rehabilitation plan	Urban rehabilitation strategies
Bring New Orleans Back commission The neighbourhood rebuilding plan The unified New Orleans plan	To develop urban design and land-use strategies on a citywide plan in order to reduce the risk of future flooding	Shifting the city's footprint and replacing the neighbourhood from the lower land level with green spaces Redevelopment of the neighbourhood plans for the flooded areas

The post-disaster urban rehabilitation models
Model of urban resilience Model of shrinking the city's footprint and land-use relocation Model of urban plan redevelopment Model of memorialization through monuments and memorials

4.1.2 Floods and Landslides in Switzerland, (2005)

The Pre-disaster Phase of the Disaster Process

Switzerland is a country exposed to many hazards such as earthquakes, floods, wildfires and windstorms. The vulnerability to these hazards grew as a result of the growing population density, increased mobility of the population and the land-use changes in these hazard-prone areas.[13] Although the floodplain areas have always been considered as a privileged area for living due to the availability of natural resources, these settlements are the ones that are most exposed to the risk of floods. Due to the population growth and urbanization, the use of the existing land and water resources became limited. This rapid urbanization was carried out in an unplanned manner, resulting in changes that increase the risk of floods. With the

[13] Lateltin et al. (2005:320)

Fig. 4.1 Photo of New Orleans before the disaster; during the disaster impact in 2005; and in the post-disaster recovery process in 2015. (Authors, August 2022)

massive influx of rural population, the urbanization process was based on the uncontrolled development of informal settlements in high-risk floodplain areas.[14] By removing the vegetation and modifying the natural soil of the slopes, the natural drainage patterns and watercourses become disrupted, decreasing the water storage

[14] WMO (2016:25)

capacity and increasing the runoff volume during rainfall.[15] In natural environments, the rainfall water is collected by the vegetation, while the land surface in urban areas that consists of urban infrastructure has less storage capacity. Land-use is defined as "the total of arrangements, activities and inputs, undertaken in a certain land cover type to produce, change or maintain it".[16] The land-use changes, such as deforestation and agricultural practices, modify the run-off generation process by reducing the infiltration capacity of the natural soil and increasing flood frequency.[17] Almost half of the Swiss forest can protect the urban areas and the urban infrastructure during events of extreme floods.[18]

The Disaster Event Phase of the Disaster Process

During the period from 21 to 23 August 2005, a continuous rainfall in the areas of northern Switzerland caused severe flooding that resulted in activation of landslide processes on numerous slopes. Six people lost their lives in the floods and landslides[19] and the total cost of the property damage was more than three billion Swiss Francs (CHF).[20] The floods enhanced many morphologic changes: the bank slopes collapsed; the erosion on the river caused the shifting of the channels that led to an overbank sedimentation; debris flows; dike failure and overflows; upwelling of ground water; and over 5000 landslides and slope-type debris flows, that were recorded in August 2005, particularly affecting the northern slopes of the Alps. During the floods, the streets became streaming canals containing a large volume of floating wood debris, out of which two thirds was the wood from the landslides. At least 110,000 m^3 of wood was mobilized.[21]

During the flood, people were evacuated to the higher ground levels and many emergency barriers were established in form of permanent and solid mobile elements. Deep fundamentals were established in the river zones together with slide measures through reinforced walls on the backside, temporary walls to keep the mud away and finally strong retaining walls for debris flows.[22]

---

[15] WMO (2016:17)
[16] WMO (2016:5)
[17] WMO (2016:1)
[18] FOEN (2015)
[19] DETEC (2008)
[20] Rickenmann et al. (2015)
[21] DETEC (2008:6)
[22] Petrascheck (2005)

The Post-disaster Phase of the Disaster Process

The post-disaster recovery plan of Switzerland consisted of a flood risk mitigation plan that would decrease the flood risk by land-use planning policies.[23] The recovery strategy involved: reducing the flood hazard, by increasing the drainage system, storage of runoff, enhancing the river capacity and dams and flood embankments; reducing the flood volume and exposure, by floodplain regulations, redevelopment of land-use policies and housing and building codes; and preserving the natural resources of floodplains, by enlarging the river space to enable space for extreme floods, soil pollution regulation and increasing the flood awareness.[24] The main strategic tools that would facilitate this awareness are the development and communication of the flood hazard maps and emergency plans built in the hazard maps for every flood-prone area. These maps show the probability of flood occurrences, the possibility of land-use and adaption of the use of the building during the flood impact.[25,26]

The Post-disaster Urban Rehabilitation Models

Based on the disaster process overview, the post-disaster rehabilitation of the floods and landslides in Switzerland is reflected through several post-disaster rehabilitation models. The emergency strategies that were used during evacuation of the people to the higher level grounds, in terms of establishing emergency barriers through temporary and retaining walls, reveal the post-disaster urban rehabilitation model of urban resilience, to cope with the disaster impact. In order to decrease the flood risk, the post-disaster rehabilitation involved measures that would increase the drainage system, enhance the river capacity and the embankments of floods, revealing the post-disaster urban rehabilitation model of reducing the flood hazard. The measures also included the floodplain regulation, redevelopment of land-use policies, housing and building codes, revealing the post-disaster rehabilitation model of reducing the flood volume and exposure. By enlarging the river space, regulating the soil pollution and increasing the flood awareness, the post-disaster recovery process reveals the post-disaster urban rehabilitation model of preserving the natural resources of the floodplain. The mapping of flood-prone areas in the post-disaster recovery process enables the development of emergency plans, which in turn increases the flood awareness, revealing the post-disaster urban rehabilitation model of raising the flood awareness (Table 4.2; Fig. 4.2).

---

[23] WMO (2016:2)

[24] WMO (2016:32)

[25] WMO (2016:35)

[26] FOEN (2015)

Table 4.2 Disaster process overview of the floods and landslides in Switzerland

<table>
<tr><td colspan="4">Disaster process overview</td></tr>
<tr><td colspan="3">Disaster category</td><td>Natural</td></tr>
<tr><td colspan="3">Disaster subcategory</td><td>Hydrological disasters: floods and landslides</td></tr>
<tr><td colspan="3">Case study disaster (country, region/city, date, year of disaster event)</td><td>Switzerland, August 21–23, 2005</td></tr>
<tr><td colspan="4">Disaster process overview—use of the urbanscape</td></tr>
<tr><td colspan="4">The pre-disaster phase—prevention and mitigation</td></tr>
<tr><td colspan="2">Urban pattern of the everyday life</td><td colspan="2">Urban vulnerability</td></tr>
<tr><td colspan="2">Switzerland is a country exposed to many hazards such as earthquakes, floods, wildfires and windstorms
Although the floodplain areas have always been considered as privileged area for living, due to the available natural resources, these settlements are the most exposed to the flood risk
With the massive influx of rural population, the urbanization process was based on the uncontrolled development of informal settlements in high-risk floodplain areas
Almost half of the Swiss forest can protect the urban areas and the urban infrastructure during extreme flood events</td><td colspan="2">The growing population density, increased mobility of the population and the land-use changes in these hazard-prone areas
By removing the vegetation and modifying the natural soil of the slopes, the natural drainage patterns and watercourses become disrupted, decreasing the water storage capacity and increasing the runoff volume during rainfall
The land-use changes, such as deforestation and agricultural practices, modify the run-off generation process, by reducing the infiltration capacity of the natural soil and increasing flood frequency</td></tr>
<tr><td colspan="4">The disaster event phase
Disaster impact—urbanscape identity disruption</td></tr>
<tr><td colspan="2">Direct disaster effect</td><td>Secondary effect</td><td>Indirect disaster effect</td></tr>
<tr><td>Population impact</td><td>Six people lost their life in the floods and landslides</td><td rowspan="4">Shifting of the channels, which led to an overbank sedimentation
Over 5000 landslides and slope-type debris flows, that were recorded in August 2005, particularly affecting the northern slopes of the Alps</td><td rowspan="4">During the floods, the streets became streaming canals containing a large volume of floating wood debris, out of which two third was the wood from the landslides. At least 110,000 m^3 of wood was mobilized</td></tr>
<tr><td>Property impact</td><td>The total cost of the property damage was more than three billion Swiss francs (CHF)</td></tr>
<tr><td>Urban impact</td><td>The floods enhanced many morphologic changes: The bank slopes collapsed; the erosion on the river caused the shifting of the channels; dike failure and overflows; and upwelling of ground water</td></tr>
<tr><td>Urban resilience</td><td>During the flood, people were evacuated to the higher ground levels, many emergency barriers were established in form of permanent and solid mobile elements</td></tr>
</table>

(continued)

Table 4.2 (continued)

The disaster event phase Disaster impact—urbanscape identity disruption			
Direct disaster effect		Secondary effect	Indirect disaster effect
	Deep fundaments were established in the river zones, slide measures through reinforced walls on the backside, temporary walls to keep the mud away and strong retaining walls for debris flows		

The post-disaster phase Post-disaster recovery—post-disaster urbanscape identity rehabilitation		
Rehabilitation program	Rehabilitation plan	Urban rehabilitation strategies
Flood risk mitigation plan	Decrease the flood risk by land-use planning policies Emergency plans build upon the hazard maps of every flood-prone area. They show the probability of flood occurrences and the possibility of land-use in terms and adaption of the use of the building during the flood impact	Reducing the flood hazard, by increasing the drainage system, storage of runoff, enhancement of the river capacity and dams and flood embankments; reducing the flood volume and exposure, by floodplain regulations, redevelopment of land-use policies and housing and building codes; and preservation of the natural resources of floodplains, by enlarging the river space to enable space for extreme floods, soil pollution regulation and increasing the flood awareness

The post-disaster urban rehabilitation models
Model of urban resilience Model of reducing the flood hazard Model of reducing the flood volume and exposure Model of preserving the natural resources of the floodplain Model of raising the flood awareness

4.1.3 *Floods in Pakistan (2010)*

The Pre-disaster Phase of the Disaster Process

According to the Global Climate Risk Index 2017, Pakistan is the seventh country in the world that is seriously affected by the impact of climate change. Over the last 70 years, the country has suffered from an increasing number of heavy rainfalls and an increase in average temperatures.[27] During the monsoon season, on July 22, 2010, Pakistan experienced a heavy monsoon rainfall that resulted in one of the most destructive floods in the country's history. Most of the people are settled in

[27] Sathar et al. (2018:5)

Fig. 4.2 Photo of Altdorf, during the floods in August 1987; during the floods in August 2005; and in the post-disaster recovery process in October 2008. (Authors, August 2022)

floodplain or flood-prone areas, due to the lack of updated maps indicating the existence of areas vulnerable to flood hazards.[28]

[28] Iqbal et al. (2014:23)

The Disaster Event Phase of the Disaster Process

The devastating rain that started on July 18 and ended on September 10[29] caused the Indus River to breach levees and overflow into and flood the northwest and east Pakistan, affecting both rural and urban population. The floods affected 18 million people,[30] causing the death of approximately 2000 people and leaving over 3000 injured.[31] During the floods, around 192,786 houses were damaged, and 728,192 houses destroyed.[32] About 20% of the country's total area was underwater. The damage to the urban infrastructure and facilities was extensive, having a negative impact on the country's economic growth. Due to the lack of clean water, the people were faced with an outbreak of diseases. Areas of flooding suffered unprecedented scales of agricultural, forestry and property losses, such as agricultural crops, livestock and seeds. More than two million hectares of standing crops were damaged or lost.[33] The flood damaged the existing the regulatory system of the Indus River basin and increased the vulnerability to future flood hazard. The urban resilience of the people affected was reflected through flight and evacuations and further followed through the disaster effects, with temporary displacements and migrations to new areas.[34] About 10.1 million people were in the need of shelter and humanitarian aid.[35]

The Post-disaster Phase of the Disaster Process

The flood response consisted of three interrelated phases: the rescue and relief phase, based on emergency operations, ensuring food and shelter; the early recovery, based on rebuilding the affected houses and restoring the live hoods; and the reconstruction and rehabilitation phase based on rebuilding the remaining damage within a period of 3–5 years.[36] One of the main recovery strategies was based on the development of geo-spatial maps, to be used as a key planning tool in disaster response. This mapping would be based on the use of satellite images in the pre-disaster phase, during the disaster event phase and in the post-disaster phase of the disaster process, through the assessment of the disaster situations, by using satellite imagery analysis

---

[29] Iqbal et al. (2014:9)

[30] Mutaza (2011:3)

[31] Kronstadt et al. (2010:1)

[32] Finance Division, Government of Pakistan (2011:221)

[33] Kronstadt et al. (2010:3)

[34] Emerson (2011)

[35] Finance Division, Government of Pakistan (2011:221)

[36] Emerson (2011:21)

and mapping work.[37] The geo-spatial maps could conduct mapping in terms of vulnerability assessment of areas vulnerable to flood hazards, based on populations density and areas vulnerable to flood hazards to develop hazard management maps; identification of possible shelter sites in the pre-disaster phase; possibility of utilizing flood water with the use of new channels for flood water diversion; assessment and analysis of the impact of flood in the agriculture areas; and in terms of identifying the abandoned river channels to develop a multi-purpose drainage system, by activating the old drainage routes.[38] Since the floods in 2010, the intensity of floods in Pakistan has only increased and during the previous years, people have been forced to live with annual impact during the monsoon season. For this reason, the fourth national flood prevention plan (2015–2025) was developed in 2015, in order to find an approach and strategy to mitigate future floods.

The Post-disaster Urban Rehabilitation Models

The disaster process overview of the floods in Pakistan reveals several post-disaster urban rehabilitation models. Since the post-disaster rehabilitation process was based on the rehabilitation strategy to assess the disaster situations, the disaster process would be analysed with the use of satellite images in the pre-disaster phase, during the disaster event phase and in the post-disaster phase, revealing the post-disaster urban rehabilitation model of developing geo-spatial maps. The development of geo-spatial maps was conducted based on the assessment of areas vulnerable to floods, revealing the post-disaster urban rehabilitation model of mapping hazard-prone areas to develop hazard management maps. The hazard management maps were based on identifying possible shelter areas in the pre-disaster phase, revealing the post-disaster urban rehabilitation model of mapping shelter areas; the possibility of utilizing flood water, revealing the post-disaster urban rehabilitation model of flood water diversion; and by activating the old drainage routes to establish a multi-purpose drainage system, revealing the post-disaster urban rehabilitation model of mapping abandoned river channels (Table 4.3; Fig. 4.3).

[37] Iqbal et al. (2014:12)

[38] Iqbal et al. (2014:31)

Table 4.3 Disaster process overview of the floods in Pakistan

Disaster process overview	
Disaster category	Natural
Disaster sub-category	Hydrological disasters: floods
Case study disaster (country, region/city, date, year of disaster event)	Pakistan, July 18–September 10, 2010

Disaster process overview—use of the urbanscape

The pre-disaster phase—prevention and mitigation

Urban pattern of the everyday life	Urban vulnerability
Pakistan is the seventh country seriously affected by the impact of climate change. Over the last 70 years, the country has suffered from an increasing number of heavy rainfalls and increasing average temperatures During the monsoon season, on July 22, 2010, Pakistan experienced a heavy monsoon rainfall, that resulted in one of the most destructive floods in the country's history	Most of the people are settled in floodplain or flood-prone areas, due to the lack of updated maps, indicating the existing areas vulnerable to flood hazards

The disaster event phase
Disaster impact—urbanscape identity disruption

Direct disaster effect		Secondary effect	Indirect disaster effect
Population impact	The floods affected 18 million people, causing the death of approximately 2000 people and leaving over 3000 injured	The lack of clean water Areas of flooding suffered unprecedented scales of agricultural, forestry and property losses, such as agricultural crops, livestock and seeds. More than two million hectares of standing crops were damaged or lost The flood damaged the existing regulatory system of the Indus River basin and increased the vulnerability to future flood hazard	The people were faced with an outbreak of diseases Temporary displacements and migrations to new areas About 10.1 million people were in the need of shelter and humanitarian aid
Property impact	Around 192,786 houses were damaged, and 728,192 houses destroyed		
Urban impact	About 20% of the country's total area was underwater. The devastating rain caused the Indus River to breach levees and overflow into flooding in the northwest and East Pakistan, affecting both rural and urban population The damage to the urban infrastructure and facilities was extensive		
Urban resilience	Reflected through flight and evacuations		

The post-disaster phase
Post-disaster recovery—post-disaster urbanscape identity rehabilitation

Rehabilitation program	Rehabilitation plan	Urban rehabilitation strategies
The flood response that consisted of three interrelated	The development of geo-spatial maps to be used as	The geo-spatial maps could conduct mapping in terms of

(continued)

Table 4.3 (continued)

The post-disaster phase Post-disaster recovery—post-disaster urbanscape identity rehabilitation		
Rehabilitation program	Rehabilitation plan	Urban rehabilitation strategies
phases: The rescue and relief phase, based on emergency operations, ensuring food and shelter The early recovery phase, based on rebuilding the affected houses and to restore the live hoods and the reconstruction and rehabilitation phase based on rebuilding the remaining damage within a period of 3–5 years	a key planning tool in disaster response This mapping would be based on the use of satellite images in the pre-disaster phase, during the disaster event phase and in the post-disaster phase of the disaster process, through the assessment of the disaster situations, by using satellite imagery analysis and mapping work	vulnerability assessment of areas vulnerable to flood hazards based on populations density and areas vulnerable to flood hazards to develop hazard management maps; identifying possible shelter sites in the pre-disaster phase; possibility of utilizing flood water with the use of new channels for flood water diversion Assessment and analysis of the impact of flood in the agriculture areas and identifying the abandoned river channels to develop a multipurpose drainage system by activating the old drainage routes In 2015, development of the fourth national flood prevention plan (2015–2025)
The post-disaster urban rehabilitation models		
Model of mapping hazard-prone areas Model of mapping shelter areas Model of mapping flood water diversion Model of mapping abandoned river channels		

Fig. 4.3 Photo of Pakistan village in Nowshera city, during the pre-disaster phase; during the disaster event on July 30, 2010; and in the post-disaster phase, reflected through repeated floods, on August 2, 2011. (Authors, August 2022)

References

DETEC (2008) The floods of 2005 in Switzerland. Synthesis report on the event analysis. In: Swiss Federal Office for the Environment. Federal Department for the Environment, transport, energy and communications. DETEC, Geneva

Emerson S (2011) Floods in Pakistan in the state of environmental migration 2010. IDDRI, Paris, pp 17–26

FEMA (2006) Hurricane Katrina in the Gulf coast. Mitigation assessment team report. Building performance observations, recommendations and technical guidance. FEMA, Washington, DC

Finance Division, Government of Pakistan (2011) Pakistan: flood impact assessment. 2 Special Section. Economic Survey 2010–2011. Finance Division, Islamabad

FOEN (2015) Living with natural hazards. In: Natural resources in Switzerland, environment. FOEN, Geneva

Iqbal I, Iqbal Z, Ravan S (2014) Effective use of space-based information to monitor disasters and it's impacts. In: Lessons learnt from floods in Pakistan. Pakistan Space and Upper Atmosphere Research Commission. SUPARCO, Islamabad

Kondo T (2008) Planning for post-disaster recovery in New Orleans after hurricane Katrina. Disaster Reduction and Human Renovation Institution, Kobe

Kronstadt A, Pervaze S, Bruce V (2010) Flooding in Pakistan: overview and issues for congress. Congressional Research Service, Washington, DC

Lateltin O, Haemmig C, Raetzo H, Bonnard C (2005) Landslide risk management in Switzerland. Landslides 2:313–320. https://doi.org/10.1007/s10346-005-0018-8

Leroy J, Amdal J, Cariolet J-M, Diab Y, Becue V (2016) Vulnerability and social resilience: comparison of the neighbourhoods in New Orleans after Hurricane Katrina. In: Paper presented at the European conference on flood risk management. EDP Sciences, Les Ulis. https://doi.org/10.1051/e3sconf/20160708002

Morris B (2009) Hurricane Katrina and the arts of remembrance. In: Moment to monument, cultural studies, vol 32. Verlag, Bielefeld, pp 155–168

Mutaza N (2011) Pakistan floods 2010. Evaluation of the relief phase of the International Federation of Red Cross Crescent Societies. Pakistan Red Crescent Society, Islamabad

NOAA (2006) Hurricane Katrina August 23–31, 2005. Service Assessment. NOAA, Silver Spring

Omićević N, Bojanić Obad Šćitaroci B (2018) The urbanscape duality during a disaster and post-disaster. Prostor 26:106–115. https://doi.org/10.31522/p.26.1(55).8

Petrascheck A (2005) Floods August 2005 in Switzerland. Federal Office for Water and Geology. United Nations International Strategy for Disaster Risk Reduction. UNISDR, Geneva

Rickenmann D, Badoux A, Hunzinger L (2015) Significance of sediment transport process during piedmont floods: the 2005 flood events in Switzerland. Earth Surf Process Landf 41:224–230. https://doi.org/10.1002/esp.3835

Sathar Z, Khalil M, Sabahat H, Maqsood S, Kiren K (2018) Climate change, resilience, and population dynamics in Pakistan: a case study of the 2010 floods in Mianwali District. The Population Council Inc., Islamabad. https://doi.org/10.31899/pgy7.1013

Saxena M (2016) Role of open spaces in disaster management. AGORA, Baltimore

WMO (2016) The role of land-use planning in flood management. Integrated flood management tools series. World Meteorological Organization, Geneva

Chapter 5
Comparative Analysis of the Selected Meteorological Disaster Process Case Studies

Abstract This chapter presents an analytical overview of the selected meteorological disaster case studies in the world that include storms, typhoons, cyclones, hurricanes, winter storms, tornadoes and tropical storms. The following disaster case studies are analysed within the meteorological disaster natural subcategory: Cyclone in Bangladesh (1991); Cyclone in Australia (2017) and Hurricane in the United States of America (2012), based on the use of the urbanscape, through its built and natural elements: the urban pattern of the everyday life and the urban vulnerability during the pre-disaster phase; the direct urban impact during the disaster event phase and the urban rehabilitation during the post-disaster phase.

Keywords Natural disasters · Meteorological disaster · Cyclone · Hurricane

5.1 Comparative Analysis of the Selected Meteorological Disaster Process Case Studies

5.1.1 Cyclone in Bangladesh, (1991)

The Pre-disaster Phase of the Disaster Process

The unique geographic location of Bangladesh and the tropical monsoon climate made the country more exposed and vulnerable to the impact of natural disasters. Because of its low elevation above the sea level, shallow continental shelf and high dense coastal urbanization,[1] the entire coast of Bangladesh is mainly affected by tropical cyclones, followed by storm surges and torrent rainfalls.[2] The Bay of Bengal, as one of the most dangerous basins of tropical cyclones in the world, is being struck by an average of 16 tropical cyclones, each year during the two periods,

[1] Shitangsu (2013:1)

[2] Shitangsu (2013:2)

N. Omićević, B. Bojanić Obad Šćitaroci, *The Urban Rehabilitation of Post-Disaster Scapes*, Disaster Resilience and Green Growth,
https://doi.org/10.1007/978-981-19-9505-7_5

April until May and October until December.[3] During the period of 1797 until 1991 Bangladesh has been affected by 59 tropical cyclones.[4] Most of the population lives in the hazard prone areas along the coast. These settlements consist of houses made of bamboo and thatch that makes them more vulnerable to the cyclone impact. Only a few live in houses that are made of brick and concrete. Despite these vulnerable facts, the population density along the coast continuously increased in the hazard prone areas.

The Disaster Event Phase of the Disaster Process

On April 29, 1991, a category 4 cyclone struck the south coast of Bangladesh, causing the death of nearly 140,000 people and severe damage.[5] The storm surge caused whole villages to be swept away. Around one million homes were destroyed or damaged, leaving ten million people homeless.[6] Around 74,000 acres of cropland were destroyed and another 300,000 acres of cropland were damaged, leading to the death of more than one million cattle. Since the fields were covered by salt water, the drinking water was contaminated. The destruction of urban infrastructure was widespread. Many bridges were washed away, and the transportation infrastructure was damaged.[7]

The previous cyclone occurrences lead to the construction of cyclone shelters in vulnerable coastal areas. After the cyclone of 1970, several mitigation measures were undertaken in order to cope with the future disaster impacts. They included the construction of multi-purpose cyclone shelters and coastal embankments of low height to protect the low-laying agricultural land, improvement of the housing conditions, and coastal afforestation with planting of seedlings on the slopes of the embankments, the social awareness and mobilization.[8] Unfortunately, during the cyclone impact in 1991, these coastal embankments couldn't resist the high storm surge and the areas without sufficient afforestation were easily damaged.

The Post-disaster Phase of the Disaster Process

In the aftermath of the disaster, in 1992, 512 cyclone shelters were established along the coast of Bangladesh that could accommodate 2500 people. For the livestock, additional 200 cyclone shelters have been distributed in the hazard prone areas that could accommodate 400 livestock.[9] The disaster management activities in the post-disaster phase consisted of a 3-year UNDP technical assistance project involving

[3] Bimal et al. (2010)
[4] Damen and van Westen (2018:10)
[5] Bimal et al. (2010:90)
[6] McCarthy (1994:2)
[7] McCarthy (1994:3)
[8] Miyan (2006:4)
[9] Bimal et al. (2010:94)

various program steps. The attention was mainly directed towards the afforestation in the coastal region, through the Program of "Development and Settlement of Accreted Land".[10] The mitigation measures involved the utilization of newly accreted land for planting mangrove plants, protection of the existing mangrove forest and replanting the degraded forests in order to reduce the water flow in case of storm surges. In November 2008, another category 4 cyclone struck the coast of Bangladesh, causing the death of 4000 people. The main factors that contributed to the minimized population impact, referred to the resilience of the Sundarbans mangrove forests[11] along the southern coast of Bangladesh, the expansion of the network of cyclone shelters and the fact that the cyclone made landfill during low tides, which minimized the storm surge impact.[12]

The Post-disaster Urban Rehabilitation Models

The disaster process overview of the cyclone in Bangladesh reveals several urban rehabilitation models. Due to the continuous impact of tropical cyclones, the country implemented several mitigation measures in 1970, in terms of the use of the urbanscape to cope with the future disaster impacts, revealing the post-disaster urban rehabilitation model of urban resilience. After the cyclone in 1991, the post-disaster recovery process began with the establishment of cyclone shelters to accommodate the people and livestock along the coastal region, revealing the post-disaster urban rehabilitation model of establishing multi-purpose cyclone shelters. The rehabilitation strategies where further implemented in the afforestation of the coastal region, revealing the post-disaster urban rehabilitation model of afforestation and aimed to reduce the water flow in case of storm surges, by the protection of the existing mangrove forest and replanting the degraded forests, revealing the post-disaster urban rehabilitation model of utilizing newly accreted land. Despite the country's vulnerability to being exposed to continuous cyclone impact, the population density along the coastal region keeps growing, revealing the need for the post-disaster rehabilitation model of raising cyclone awareness (Table 5.1; Fig. 5.1).

5.1.2 Cyclone in Australia, (2017)

The Pre-disaster Phase of the Disaster Process

The region of Australia, especially the coast of Queensland, is exposed to tropical cyclones that occur in November and throughout April. From 1961 until 2017, the region was struck by an average of 12 cyclone occurrences per year, where five of

---

[10]Miyan (2006:6)

[11]The Sundarbans mangrove forest is of the largest such forests in the world, that lies on the delta of the Ganges, Brahmaputra and Meghna rivers on the Bay of Bengal

[12]Bimal et al. (2010:91)

Table 5.1 Disaster process overview of the cyclone in Bangladesh

<table>
<tr><th colspan="4">Disaster process overview</th></tr>
<tr><td colspan="3">Disaster category</td><td>Natural</td></tr>
<tr><td colspan="3">Disaster sub-category</td><td>Meteorological disasters: cyclone</td></tr>
<tr><td colspan="3">Case study disaster (country, region/city, date, year of disaster event)</td><td>Bangladesh, April 29–30, 1991</td></tr>
<tr><td colspan="4">Disaster process overview—use of the urbanscape</td></tr>
<tr><td colspan="4">The pre-disaster phase—prevention and mitigation</td></tr>
<tr><td colspan="2">Urban pattern of the everyday life</td><td colspan="2">Urban vulnerability</td></tr>
<tr><td colspan="2">The Bay of Bengal, as one of the most dangerous basins of tropical cyclones in the world, is being struck by an average of 16 tropical cyclones every year during the two periods, April until May and October until December</td><td colspan="2">Low elevation above the sea level, shallow continental shelf and high dense coastal urbanization
Most of the population lives in the hazard prone areas along the coast in houses made of bamboo and thatch that makes them more vulnerable to the cyclone impact. Only a few live in houses that are made of brick and concrete
Continuously increasing population density in the hazard prone areas</td></tr>
<tr><td colspan="4">The disaster event phase—disaster impact—urbanscape identity disruption</td></tr>
<tr><td colspan="2">Direct disaster effect</td><td>Secondary effect</td><td>Indirect disaster effect</td></tr>
<tr><td>Population impact</td><td>Nearly 140,000 people died</td><td rowspan="4">Storm surges and torrent rainfalls</td><td rowspan="4">The property destruction left ten million people homeless
Since the fields got covered by salt water, the drinking water got contaminated
In the aftermath of the disaster, in 1992, 512 cyclone shelters were established along the coast of Bangladesh that could accommodate 2500 people. For the livestock, additional 200 cyclone shelters have been distributed in the hazard prone areas that could accommodate 400 livestock</td></tr>
<tr><td>Property impact</td><td>Around one million homes were destroyed or damaged
Around 74,000 acres of cropland were destroyed and another 300,000 acres of cropland were damaged, leading to the death of more than one million cattle</td></tr>
<tr><td>Urban impact</td><td>The coastal embankments could not resist the high storm surge and the areas without sufficient afforestation got easily damaged
The destruction of urban infrastructure was widespread. Many bridges got washed away and the transportation infrastructure was damaged</td></tr>
<tr><td>Urban resilience</td><td>The construction of multi-purpose cyclone shelters and coastal embankments of low height to protect the low-laying agricultural land, improvement of the housing conditions, coastal afforestation with planting of seedlings on the slopes of</td></tr>
</table>

(continued)

Table 5.1 (continued)

The disaster event phase—disaster impact—urbanscape identity disruption			
Direct disaster effect		Secondary effect	Indirect disaster effect
	the embankments, the social awareness and mobilization		

The post-disaster phase Post-disaster recovery—post-disaster urbanscape identity rehabilitation		
Rehabilitation program	Rehabilitation plan	Urban rehabilitation strategies
A 3-year UNDP technical assistance project that involved various program steps—program of "development and settlement of accreted land"	Afforestation in the coastal region	The mitigation measures involved the utilization of newly accreted land for planting mangrove plants, protection of the existing mangrove forest and replanting the degraded forests in order to reduce the water flow in case of storm surges

The post-disaster urban rehabilitation models
Model of urban resilience Model of construction of multi-purpose cyclone shelters Model of afforestation Model of utilizing newly accreted land Model of raising cyclone awareness

them make landfall, causing destructive winds and heavy rains.[13] Despite these frequent disaster occurrences, the Australian population has little awareness of the climate change effects. As a result of the continuous rain events during the period of 2010 and 2011, more than 200,000 people were displaced. The floods affected more than 78% of Queensland and caused the death of 33 people and severe property and environmental damage. The scale of the disaster impact in 2010 and 2011 led to the establishment of the Queensland Flood Commission that would serve as a platform for the development of future disaster mitigation strategies, before, during and after the disaster, in order to raise the population's awareness. Despite the known vulnerability risks, the cyclone that struck the Queensland coast in 2017 caused long-term and widespread impacts of the entire region.[14]

The Disaster Event Phase of the Disaster Process

On March 27, 2017, the severe tropical cyclone "Debbie" made landfall on the northern Queensland coast of Australia, as a Category 4 system[15] and lasted until

[13] Mortlock et al. (2019:2604)

[14] Gallina (2017:10)

[15] Category 4 tropical cyclone of very destructive winds with gusts of 225–279 km/h, causes significant roofing loss and structural damage

Fig. 5.1 Photo of aerial view of Bangladesh during the cyclone disaster in 1970; during the cyclone disaster impact in 1991; and in the post-disaster recovery process in 2019, struck again by floodwaters. (Authors, August 2022)

March 31, 2017. With torrential rainfall and the highest wind records of gusts up to 265 km/h, the cyclone caused widespread flooding across Queensland, causing severe land and infrastructure damage, destroying homes, crops and leading to the death of 12 people.[16] This widespread flooding occurred in North Queensland, then emerged to rapid on-set flooding in the south-east region of Queensland and finished with slow on-set flooding in Central Queensland.[17] After 24 h being inland, the flooding spread across the entire region leading to severe environmental, economic and social destruction. Around 3300 properties in Queensland were damaged, leaving 944 properties completely inhabitable.[18] The tourism industry was heavily impacted, roads, railways and bridges were damaged or blocked by fallen trees. The airport and seaport were closed. Out of 1040 natural preserved areas in Queensland, 223 areas of protected land were affected by the floods.[19] The flooding resulted in landslides, fallen vegetation and erosion. A total of 73 landslides were registered in the Lamington Nation Park Road alone. A total of 223 parks and forests within the Queensland coast were affected by the cyclone.[20] Thousands of homes were left without electricity and water supply. During the flood impact, resilience was established through strategic evacuation of population led by the Queensland evacuation management through five stages: the decision, warning, withdrawal, shelter and return.[21]

The Post-disaster Phase of the Disaster Process

In the post-disaster phase of the flood impact, Queensland recovery was based on the development of the "State Recovery Plan 2017–2019—Operation Queensland Recovery", with the aim to "recover, reconnect and rebuild more resilient communities" using five functional fields: "human and social, economic, environmental, roads and infrastructure and property".[22] In environmental terms, the plan encompasses the environmental recovery package that consists of several rehabilitation strategies: the coastal removal of debris, restoration of mangrove and other coastal vegetation and restoration of the beaches; clean-up and restoration of green waste; soil conservation and improvement in the field of mapping and collecting data.[23] According to the Debbie Recovery Progress Report, 8 months after the cyclone impact, all parks and forest were reopened or partially opened and all fire breaks have been repaired. The damaged roads were reopened or provided alternatives, and according to the transport reconstruction plan, all urban infrastructure reconstructions will be completed by June 2019.

---

[16] Lenzen et al. (2019:3)

[17] Operation Queensland Recovery (2017:4)

[18] Operation Queensland Recovery (2017:4)

[19] Operation Queensland Recovery (2017:25)

[20] NPAQ (2017:5)

[21] Operation Queensland Recovery (2017:4)

[22] Operation Queensland Recovery (2017:5)

[23] Operation Queensland Recovery (2017:26)

The Post-disaster Urban Rehabilitation Models

Based on the disaster process overview, the post-disaster rehabilitation of the cyclone in Australia is reflected through several post-disaster urban rehabilitation models. Although the Queensland coastline is highly exposed to tropical cyclones, the population has very little awareness of the climate change effects. Due to the scale of the flood impact that occurred in 2010 and 2011, causing severe property and environmental damage, the Queensland Flood Commission was established with the main aim to develop mitigation strategies during the entire disaster process and to raise the population awareness for possible disaster occurrences, revealing the need for the post-disaster urban rehabilitation model of raising the cyclone awareness. The floods and landslides in 2017 required a strategic evacuation of the population affected, which was led by the Queensland evacuation management through five stages, revealing the post-disaster urban rehabilitation model of urban resilience. Based on the post-disaster Queensland Recovery Plan, the environmental rehabilitation consisted of the coastal debris removal, conservation of soil and the restoration of coastal vegetation, green waste and beaches, revealing the post-disaster urban rehabilitation model of coastal restoration (Table 5.2; Fig. 5.2).

5.1.3 Hurricane in the United States of America, (2012)

The Pre-disaster Phase of the Disaster Process

The north-eastern coastal region of the United States is highly exposed to the impact of tropical cyclones. Since 1842 there have been 98 registered tropical cyclones, where 33 of them were hurricanes.[24] Mainly tropical cyclones that pass the north-eastern coast of the United States follow the same pattern, which is parallel to the coast or towards the East, into the Atlantic, with the exception of hurricane "Sandy" that made a westward track and landfall in October, 2012.[25] Although the coastal regions are the most vulnerable to the impact of tropical cyclones and floods, they are still inviting for living settlements. This population growth resulted in coastal urbanization and land conversion that increases the vulnerability of these coastal communities to future cyclone impacts. In 1983, the city of New York alone had 218,000 people living in flood-prone areas and 35,000 settlements. Despite the frequent disaster occurrences, this number increased by 2012, with 67,000 settlements and 398,000 people living in flood-prone areas.[26] The preparation for the hurricane impact was reflected through several actions, by providing food and water supply, generators and preparation of medical teams and medical supply.

---

[24] Torres et al. (2015:3)

[25] Torres et al. (2015:5)

[26] Torres et al. (2015:13)

Table 5.2 Disaster process overview of the cyclone in Australia

Disaster process overview	
Disaster category	Natural
Disaster sub-category	Meteorological disasters: cyclone
Case study disaster (country, region/city, date, year of disaster event)	Queensland, Australia, March 28–31, 2017
Disaster process overview—use of the urbanscape	
The pre-disaster phase—prevention and mitigation	
Urban pattern of the everyday life	Urban vulnerability
The region of Australia, especially the coast of Queensland is exposed to tropical cyclones that occur in November and last throughout April The scale of the previous disaster impacts led to the establishment of the Queensland flood commission that would serve as a platform for the development of future disaster mitigation strategies, before, during and after the disaster, in order to raise the population awareness	From 1961 until 2017, the region was struck by an average of 12 cyclone occurrences per year, out of which five make landfall, causing destructive winds and heavy rains Despite the frequent disaster occurrences, the Australian population has little awareness of the climate change effects

The disaster event phase—disaster impact—urbanscape identity disruption			
Direct disaster effect		Secondary effect	Indirect disaster effect
Population impact	12 people died	Torrential rainfall and high wind records of gusts up to 265 km/h caused widespread flooding across Queensland This widespread flooding occurred in North Queensland, then emerged to rapid on-set flooding in south-East Queensland and finished with slow on-set flooding in Central Queensland	Severe environmental, economic and social destruction The flooding resulted in landslides, fallen vegetation and erosion. In the lamington Nation Park road alone there were 73 landslides registered Thousands of homes were left without electricity and water supply
Property impact	Around 3300 properties in Queensland were damaged, leaving 944 properties completely inhabitable		
Urban impact	Roads, railways and bridges were damaged or blocked by fallen trees Out of 1040 natural preserved areas in Queensland, 223 areas of protected land were affected by the floods A total of 223 parks and forests within the Queensland coast were affected by the cyclone		
Urban resilience	Strategic evacuation of population		

The post-disaster phase—post-disaster recovery—post-disaster urbanscape identity rehabilitation		
Rehabilitation program	Rehabilitation plan	Urban rehabilitation strategies
State Recovery Plan 2017–2019—Operation Queensland Recovery	To recover, reconnect and rebuild more resilient communities, in terms of the human and social, economic, environmental, roads and infrastructure and property aspect	In environmental terms, the plan encompasses several rehabilitation strategies: The coastal removal of debris, restoration of mangrove and other coastal vegetation and restoration of the

(continued)

Table 5.2 (continued)

The post-disaster phase—post-disaster recovery—post-disaster urbanscape identity rehabilitation		
Rehabilitation program	Rehabilitation plan	Urban rehabilitation strategies
		beaches; clean-up and restoration of green waste; soil conservation and improvement in the field of mapping and collecting data Eight months after the cyclone impact, all parks and forest have reopened or partially opened and all fire breaks have been repaired. The damaged roads have been reopened or provided alternatives. According to the transport reconstruction plan, all urban infrastructure reconstructions will be completed by June 2019
The post-disaster urban rehabilitation models		
Model of raising cyclone awareness Model of urban resilience Model of coastal restoration		

The Disaster Event Phase of the Disaster Process

Hurricane Sandy made landfall on October 29, 2012 in southern New Jersey. It struck the most populated cities of New York, New Jersey and Connecticut along the Eastern coast with heavy rain, strong winds and record storm surges, followed with heavy snow fall in some regions.[27] The storm impact was devastating, more than 650,000 houses[28] were damaged or destroyed and at least 162 people died.[29] Around 70% of the buildings damaged were low-rise and built with wood.[30] The extent of the land damage was unprecedented. The storm flooded coastal roadways, tunnels, leaving more than half of the nation's daily transportation system without service. This contributed to fuel shortage along the East coast and leaving 8.5 million people without electricity supply. The storm surge caused dune and beach erosion, island breaching and inundation of wetlands, increasing the contamination of drinking water.[31] Many houses experienced fires, since the storm surge flushed streets and houses with salt water that conducts electricity. The region suffered considerable

[27] FEMA (2013)

[28] Blake et al. (2013)

[29] FEMA (2013:1)

[30] Molfetas (2017)

[31] Donovan (2013)

Fig. 5.2 Photo of aerial view of Queensland during the floods in 2010–2011; during the cyclone disaster impact in 2017; and in the post-disaster recovery process in 2019, struck again by floodwaters. (Authors, August 2022)

landscape changes, in terms of topography and vegetation loss. When the hurricane made landfall, the sea level has risen one foot and increased the vulnerability to future storm impacts.[32] In the aftermath of the storm, tens of thousands people were displaced and more than 23,000 people seek temporary shelter.[33]

The Post-disaster Phase of the Disaster Process

The post-disaster rehabilitation strategies were reflected through housing reconstruction, restoration of the energy and urban infrastructure and development of coastline resilient projects, that include rebuilding beaches and board walls and increasing coastal resilience to storm surges, by building up dunes and wetlands.[34] In line with these strategies, boardwalks were relocated to higher and less flood-prone areas and anchored into the ground to avoid floating during future floods. The beach bath houses were replaced by portable structures. Beach parking lots have been resurfaced with native materials that can be reused and do not leave asphalt debris on the beach in case of flooding. The post-disaster recovery has been seen as an opportunity to replace the damaged structures with resilient alternative and adapt to the climate change effects.[35]

The Post-disaster Urban Rehabilitation Models

The disaster process overview of the hurricane impact in the United States of America, north-eastern coastline, reveals several post-disaster urban rehabilitation models. Although this region is under impact of frequent disaster occurrences, its vulnerability is strongly affected by the population growth in the coastal areas. Despite the coastal vulnerability and exposure to possible floods, the number of the population living in flood-prone areas has only increased over the years, revealing the need for post-disaster rehabilitation model of raising the flood awareness. The post-disaster recovery process of the hurricane impact consisted of housing reconstruction and the restoration of the energy and urban infrastructure, revealing the post-disaster urban rehabilitation model of urban reconstruction. Along with the urban reconstruction, the rebuilding of beaches and board walls and increasing the coastal resilience to storm surges reveals the post-disaster urban rehabilitation model of building urban resilience. These rehabilitation strategies were seen as an oppor-

[32] Molfetas (2017:5)
[33] FEMA (2013:6)
[34] Molfetas (2017:6)
[35] Babson et al. (2016:126–127)

tunity to replace the damage with resilient alternatives, as a form of adaptation, rather than seeking alternatives for disaster prevention, revealing the post-disaster urban rehabilitation model of adaptation to change (Table 5.3; Fig. 5.3).

Table 5.3 Disaster process overview of the hurricane in the United States of America

Disaster process overview	
Disaster category	Natural
Disaster subcategory	Meteorological disasters: Hurricane
Case study disaster (country, region/city, date, year of disaster event)	North-eastern coastal region, United States of America, March 29, 2012
Disaster process overview—use of the urbanscape	
The pre-disaster phase—prevention and mitigation	

Urban pattern of the everyday life	Urban vulnerability
The north-eastern coastal region of the United States is highly exposed to the impact of tropical cyclones The preparation for the hurricane impact was reflected through several actions, by providing food and water supply, generators and preparation of medical teams and medical supply	The population growth in the coastal region resulted in coastal urbanization and land conversion that increases the vulnerability of these coastal communities to future cyclone impacts Around 70% of the buildings damaged were low-rise and built with wood

The disaster event phase
Disaster impact—urbanscape identity disruption

Direct disaster effect		Secondary effect	Indirect disaster effect
Population impact	162 people died	Heavy rain, strong winds and record storm surges, followed with heavy snow fall in some regions	Many houses experienced fires, since the storm surge flushed street and houses with salt water that conducts electricity The region suffered considerable landscape changes, in terms of topography and vegetation loss. When the hurricane made landfall, the sea level has risen one foot and increased the vulnerability to future storm impacts
Property impact	More than 650,000 houses were damaged or destroyed		
Urban impact	The storm flooded coastal roadways, tunnels, leaving more than half of the nation's daily transport system without service The storm surge caused dune and beach erosion, island breaching and inundation of wetlands, increasing the contamination of drinking water		

(continued)

Table 5.3 (continued)

The disaster event phase Disaster impact—urbanscape identity disruption			
Direct disaster effect		Secondary effect	Indirect disaster effect
Urban resilience	Although the coastal regions are the most vulnerable to the impact of tropical cyclones and floods, they are still tempting for living. In the aftermath of the storm, tens of thousands people were displaced and more than 23,000 people seek for temporary shelter		

The post-disaster phase Post-disaster recovery—post-disaster urbanscape identity rehabilitation		
Rehabilitation program	Rehabilitation plan	Urban rehabilitation strategies
Rebuilding plan	Coastal adaptation strategies	Housing reconstruction, restoration of the energy and transportation infrastructure, and developing coastline resilient projects that include rebuilding beaches and board walls; and increasing coastal resilience to storm surges, by building up dunes and wetlands

The post-disaster urban rehabilitation models
Model of raising flood awareness Model of urban reconstruction Model of building coastal resilience Model of adaptation to change

Fig. 5.3 Photo of aerial view of the New Jersey coast before the storm in 2011, during the storm impact in 2012; and in the post-disaster recovery process in 2017. (Authors, August 2022)

References

Babson A, Beavers R, Eissenberg M, Foley M, Hudson T, Schupp C (2016) Lessons learned from hurricane Sandy. In: Chapter 9 in coastal adaptation strategies handbook. U.S. Department of the Interior, Washington DC, pp 121–138

Bimal P, Rashid H, Shahidul I, Hunt M (2010) Cyclone evacuation in Bangladesh: tropical cyclone Gorky (1991) vs. Sidr (2007). Environ Hazards 9:89–101. https://doi.org/10.3763/ehaz.2010.SI04

Blake ES, Kimberlain TB, Berg RJ, Cangialosi JP, Beven JL (2013) Tropical cyclone report hurricane Sandy. National Hurricane Center, Miami

Damen MCJ, van Westen CJ (2018) Modelling cyclone Hazard in Bangladesh, in applications guide. In: Chap. 3: Integrated land and water information system. ILWIS: International Institute for Aerospace Survey and Earth Sciences, Enschede

Donovan S (2013) Hurricane Sandy rebuilding strategy, stronger communities, a resilient region. Hurricane Sandy Rebuilding Task Force, Washington DC. https://doi.org/10.7282/T39Z933J

FEMA (2013) Hurricane Sandy. FEMA after-action report. FEMA, Washington, DC

Gallina J (2017) Reducing the impacts of cyclones through technology: a case study of cyclone Debbie and the sunshine coast disaster hub. Aust J Emerg Manag 32(3):10–11

Lenzen M, Malik A, Kenway S, Daniels P, Lam KL, Geschke A (2019) Economic damage and spillovers from a tropical cyclone. Nat Hazards Earth Syst Sci 19(1):137–151. https://doi.org/10.5194/nhess-19-137-2019

McCarthy P (1994) Operation sea angel, a case study. RAND Corporation, Santa Monica

Miyan MA (2006) Cyclone disaster mitigation in Bangladesh. In: Regional technical workshop on "Coastal Protection" in the aftermath of the Indian Ocean Tsunami: what role for forest and trees? Khao Lak, Thailand, August 28–31, 2006. FAO Regional Office for Asia and the Pacific, Bangkok

Molfetas M (2017) Hurricane Sandy: a Retrospective. Impact Human, Brooklyn, pp 1–12

Mortlock TR, Metters D, Soderholm J, Maher J, Lee SB, Boughton G, Stewart N, Zavadil E, Goodwin ID (2019) Extreme water levels, waves and coastal impacts during a severe tropical cyclone in Northeastern Australia: a case study for cross-sector data sharing. Nat Hazards Earth Syst Sci 19(10):2067–2077. https://doi.org/10.5194/nhess-19-2067-2019

NPAQ (2017) Cyclone Debbie's impact on parks. Protected, vol 15. NPAQ, Brisbane, pp 1–15

Operation Queensland Recovery (2017) Severe tropical cyclone Debbie, 8 month progress report. In: State recovery plan 2017–2019. Operation Queensland recovery. Queensland Government, Brisbane

Shitangsu P (2013) Post-cyclone livelihood status and strategies in coastal Bangladesh. J Life Earth Agric Sci 41:1–20. https://doi.org/10.3329/rujleas.v41i0.21623

Torres A, Lowden B, Ito K, Shuman S (2015) Lessons from hurricane Sandy: analysis of vulnerabilities exposed by hurricane sandy and potential mitigation strategies in the Northeastern United States. University of Michigan, Ann Arbor

Chapter 6
Comparative Analysis of the Selected Geophysical Disaster Process Case Studies

Abstract This chapter presents an analytical overview of the selected geophysical disaster case studies in the world that include earthquakes, volcanic eruptions and tsunamis. Within the geophysical disaster natural subcategory, the following post-disaster case studies are analysed: Volcanic eruption of Mt. Pinatubo in the Philippines (1991); Earthquake and Tsunami in Japan (2011) and Earthquake and Landslides in Nepal (2015), based on the use of the urbanscape, through its built and natural elements: the urban pattern of the everyday life and the urban vulnerability during the pre-disaster phase; the direct urban impact during the disaster event phase and the urban rehabilitation during the post-disaster phase.

Keywords Natural disasters · Geophysical disasters · Volcanic eruption · Earthquake · Tsunami

6.1 Comparative Analysis of the Selected Geophysical Disaster Process Case Studies

6.1.1 Volcanic Eruption of Mt. Pinatubo in the Philippines (1991)

The Pre-disaster Phase of the Disaster Process

As the second largest archipelago in the world with over 7000 islands, the Philippines is considered as the world's main tropical cyclone generator, that induces floods, landslides and storm surges. Positioned on the edge of the Pacific Ring of Fire,[1] it is often under the impact of earthquakes and volcanic eruptions. There are about 220 volcanoes in the Philippines, out of which 22 are considered to be active.

[1] Area within the basin of the Pacific Ocean where many earthquakes and volcanic eruptions occur

N. Omićević, B. Bojanić Obad Šćitaroci, *The Urban Rehabilitation of Post-Disaster Scapes*, Disaster Resilience and Green Growth,
https://doi.org/10.1007/978-981-19-9505-7_6

All these vulnerability factors make the Philippines the fourth most disaster-prone country in the world.[2]

The Disaster Event Phase of the Disaster Process

Mount Pinatubo is located on the main island of Luzon in the Philippines and is considered to be one of the active volcanoes within the country. After more than 500 years of rest, in April 1991, the volcano started to show the first signs of a possible eruption and on June 15, 1991 Mount Pinatubo erupted. It was considered the largest volcanic eruption of the century that affected a populated area.[3] The volcano ejected massive volcanic materials, creating a cloud of volcanic ash into the air, blowing in all directions. This led to the dispersion of nearly 20 million tons of sulphur dioxide (SO_2) into the stratosphere, causing a drop in the global temperature during the period form 1991 until 1993.[4] During the eruption, 657 people died, 184 were injured and 23 were declared missing. The eruption induced heavy rains that washed the deposits of the volcanic ash and debris from the slopes into fast flowing mudflows or lahars. Since lahars streams are faster than clear water streams, their impact is more destructive. During the period from 1991 until 1997, they destroyed and buried hundreds of square miles of land causing a more severe destruction than the direct impact itself. During the eruption impact in 1991, 4979 houses were completely destroyed and 70,257 houses were damaged. Due to the disaster effects, derived from the impact, this number changed in 1992, where additional 3281 houses got destroyed and 3137 damaged.[5] The impact of lahars and floods caused an immense damage to the natural environment by destroying endemic species of flora and fauna, destructing the forest land that was covered by ash falls, and altering the landscape and land use. Around 19,799 hectares of plantations, 10,206 hectares of agro-forestry land and 96,200 hectares of agricultural land were destroyed.[6] The impact led to a widespread damage to crops and livestock, urban infrastructure, water and electricity supply. In the aftermath of the disaster impact, 11 unplanned resettlement centres were created for the people affected.[7]

The Post-disaster Phase of the Disaster Process

The post-disaster recovery process was based on the rehabilitation of the areas affected by the eruption, that was not limited only to the reconstruction and

---

[2] Eleazar (2010:67)
[3] Guzman (2004)
[4] Guzman (2004:1)
[5] Guzman (2004:3)
[6] Guzman (2004:4)
[7] Gaillard and Le Masson (2007:313)

restoration of the destroyed or damaged areas, it also initially addressed the existing vulnerabilities in order to mitigate the future disaster impacts. The first population group affected by the eruption was the Aeta population,[8] living on the slopes of the mountains. For this reason, the post-disaster rehabilitation was reflected through the resettlement of the population groups, the Aeta highlanders and the displaced lowlanders, provision of live hood programs focused on the agriculture for the displaces farmers; provision of basic social services and infrastructure; environmental management through the land use and replanning strategies.[9] The resettlement process had to take into the account the cultural diversity of the native population groups and their adaption to the imposed changes. Most of the resettlement sites did not provide sufficient space, basic facilities and services were positioned in an environment that differs from the one in the pre-disaster phase.[10] Even 16 years after the volcanic eruption, the lahars and flood continued to threaten the rehabilitation process.[11]

The Post-disaster Urban Rehabilitation Models

The disaster process overview of the volcanic eruption of Mt. Pinatubo in the Philippines reveals several post-disaster urban rehabilitation models. Addressing the existing vulnerability during the post-disaster rehabilitation was initially reflected through the resettlement of the native Aeta population, revealing the post-disaster urban rehabilitation process of population resettlement. The resettlement process was focused on the restoration of the agriculture land, social and urban infrastructure services for the displaced population, revealing the post-disaster urban rehabilitation model of resuming the pre-disaster way of life. Since the pre-disaster way of life and the cultural diversity of the Aeta population differed from the life in the resettled sites, the population was forced to adapt to the new environment and everyday life, revealing the post-disaster urban rehabilitation model of adaptation to the imposed changes. Although many years passed since the volcanic eruption, the rehabilitation process of the Philippines is being continuously disrupted with the impact of floods and lahars, revealing the post-disaster urban rehabilitation model of continuous disaster impact (Table 6.1; Fig. 6.1).

---

[8] Native people who live in scattered, isolated mountainous parts of the island of Luzon, the Philippines

[9] Guzman (2004:13)

[10] Eleazar (2010:90)

[11] Eleazar (2010:81)

Table 6.1 Disaster process overview of the volcanic eruption of Mt. Pinatubo in the Philippines

Disaster process overview			
Disaster category		Natural	
Disaster sub-category		Geophysical disasters: volcanic eruption	
Case study disaster (country, region/city, date, year of disaster event)		Mount Pinatubo, Philippines, June 15, 1991	
Disaster process overview—use of the urbanscape			
The pre-disaster phase Prevention and mitigation			
Urban pattern of the everyday life		Urban vulnerability	
The Philippines is considered the world's main tropical cyclone generator that induces floods, landslides and storm surges Positioned on the edge of the Pacific ring of fire, it is often under the impact of earthquakes and volcanic eruptions		The fourth most disaster-prone country in the world.	
The disaster event phase Disaster impact—urbanscape identity disruption			
Direct disaster effect		Secondary effect	Indirect disaster effect
Population impact	657 people died, 184 injured and 23 declared missing	The dispersion of nearly 20 million tons of sulphur dioxide (SO_2) into the stratosphere, causing the global temperature to drop during the period form 1991 until 1999	The eruption induced heavy rains that washed the deposits of the volcanic ash and debris from the slopes into fast flowing mudflows or lahars From 1991 until 1997, lahars have destroyed and buried hundreds of square miles of land, causing a more severe destruction than the direct impact itself Immense damage to the natural environment, by destroying endemic species of flora and fauna, destruction of the forest land that got covered by ash falls and the alteration of landscape and land use
Property impact	During the eruption impact in 1991, 4979 houses were completely destroyed and 70,257 houses were damaged. Due to the disaster effects, derived from the impact, this number changed in 1992, where additional 3281 houses got destroyed and 3137 damaged		
Urban impact	Around 19,799 hectares of plantations, 10,206 hectares of agro-forestry land and 96,200 hectares of agricultural land were destroyed The impact led to a widespread damage to crops and livestock and urban infrastructure		
Urban resilience	In the aftermath of the disaster impact, 11 unplanned resettlement centres were created for the people affected		

(continued)

Table 6.1 (continued)

The post-disaster phase Post-disaster recovery—post-disaster urbanscape identity rehabilitation		
Rehabilitation program	Rehabilitation plan	Urban rehabilitation strategies
The rehabilitation process of the areas affected by the eruption	Not only limited to the reconstruction and restoration of the destroyed or damaged areas, but also focused on addressing the existing vulnerabilities in order to mitigate the future disaster impacts	The resettlement of the population groups, the Aeta highlanders and the displaced lowlanders Providing live hood programs focuses on the agriculture for the displaced farmers Providing basic social services and infrastructure Environmental management through the land use and replanning strategies
The post-disaster urban rehabilitation models		
Model of population resettlement Model of resuming the pre-disaster way of life Model of adaptation to the imposed changes Model of continuous disaster impact		

6.1.2 *Earthquake and Tsunami in Japan, (2011)*

The Pre-disaster Phase of the Disaster Process

Being positioned on the intersection of at least three tectonic plates brings Japan under continuous threat of earthquakes. This makes the Pacific coastline of the Tohoku region highly vulnerable to the impact of tsunamis. Records of earthquakes that occurred in the past led to the development of resilience measures and strategies in order to mitigate future disaster impacts.[12] These measures were related to the construction of seawalls and flood gates; urban planning focused on the resilience; and evacuation planning.[13] The main pre-disaster factors that increased the vulnerability during the disaster impact were the scale, density and construction of the existing urban development, by examining if the existing buildings are constructed on built or natural earth, if the basements are occupied with parking spaces and if the density of the existing buildings limits the empty spaces within the urban structure.[14]

[12] OECD (2006)

[13] Koshimura and Nobuo (2015)

[14] Saxena (2016)

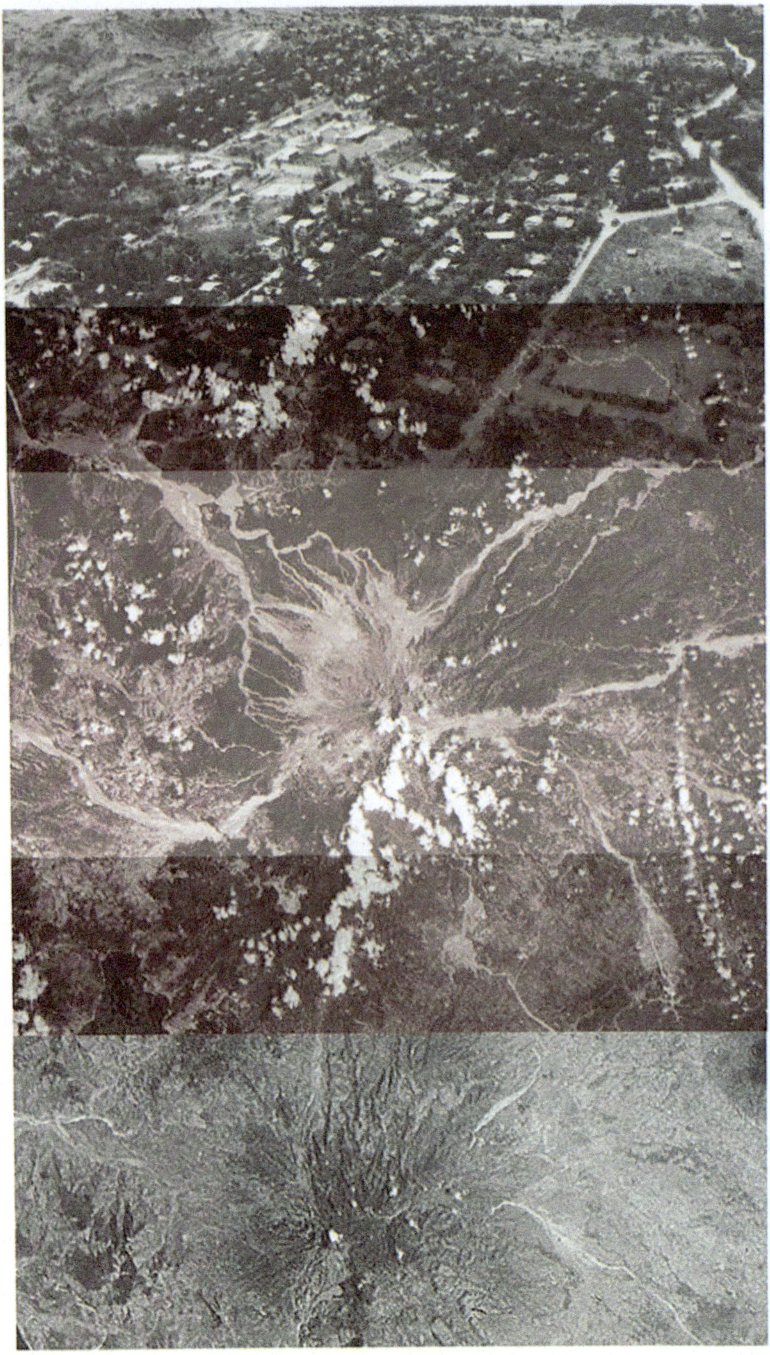

Fig. 6.1 Photo of aerial view of Mt. Pinatubo in March 1982 during the pre-disaster phase; at the aftermath of the disaster in August 1992; and the aerial view of the Aeta resettlement centre in the post-disaster recovery phase. (Authors, August, 2022)

The Disaster Event Phase of the Disaster Process

The largest recorded disaster in Japan was the earthquake in 2011, along the Pacific side of the Tohoku region, followed with a massive tsunami. According to the damage statistics, 6 months after the earthquake, the combined impact of the earthquake and tsunami left a total of 15,879 deaths and 4086 missing persons. The property damage was divided into the destruction of houses by seismic motion and the houses along the coast that got washed away. As of 6 months after the earthquake, 125,530 buildings were completely destroyed and 240,332 partially damaged.[15] In terms of the landscape impact, the tsunami led to the flooding of agricultural land ruining 23,600 hectares of farmland and altering the coastline with the 0.6 m subsidence of 400 km of Japan's east coast.[16] The impact on the urban infrastructure was reflected through the damage of many road facilities, especially the main route along the coast, damages to bridge bearings and joints, the water drainage systems and electrical power station and electricity transmission facilities. Due to the tsunami impact, many bridge decks and road embankments were washed away. The nuclear disaster in the Fukushima Prefecture, that followed the disaster, leads to a long-term evacuation. Because of the level of radiation, many people were forced to leave their homes and 335,000 refugees were in need of essential supplies for survival.[17] In order to mitigate the easy evacuation of people during the disaster impact, the need of having open public spaces was essential. The open public spaces during the disaster impact enabled evacuation of people. Many green spaces were occupied with shelters. The urban pattern of the everyday life that enhances the movement within the existing roads influenced the evacuation routes during the disaster impact.[18]

The Post-disaster Phase of the Disaster Process

The post-disaster rehabilitation process consisted of the establishment of the National Reconstruction Design Council, composed of 15 members, coming from different professional fields. The main aim of this Council was to develop a concept and strategy for long-term recovery. The Council's report consisted of four rehabilitation programs, proposing future land-use patterns. These future urban patterns would reduce the risk and vulnerability of tsunamis. These strategies consisted of land-use relocation, in terms of relocating the communities away from the zones with the risk of tsunami hazard, in "The Collective Relocation Program"; raising the base elevation of land and the height of the flood bank in "The Land Readjustment Program"; enabling the reconstruction of housing property, in "The Public Housing

[15] Kazama and Noda (2012:781–782)

[16] Norio et al. (2011:36)

[17] Norio et al. (2011:35)

[18] Omićević and Bojanić Obad Šćitaroci (2018)

Program" and the redevelopment of urban infrastructure and facilities in the disaster aftermath, in "The Tsunami Recovery Zone Program".[19]

The Post-disaster Rehabilitation Models

Based on the disaster process overview, the post-disaster rehabilitation of the earthquake and tsunami in Japan is reflected through several post-disaster urban rehabilitation models. In terms of long-term recovery, the post-disaster urban rehabilitation process consisted of developing future land-use pattern. This was reflected through relocating the communities away from the zones with the risk of tsunami hazard, revealing the post-disaster urban rehabilitation model of land-use relocation. Raising the base elevation of land and the height of the flood bank in the aftermath of the disaster reveals the post-disaster urban rehabilitation model of land readjustment. The severe damage of the earthquake and tsunami impact required the reconstruction of housing property and the redevelopment of urban infrastructure, revealing the post-disaster urban rehabilitation model of urban reconstruction (Table 6.2; Fig. 6.2).

6.1.3 Earthquake and Landslides in Nepal, (2015)

The Pre-disaster Phase of the Disaster Process

Nepal is considered a high disaster-prone country, with a significant record of 19 major earthquakes since the twelfth century. The capital city Kathmandu is the most vulnerable city in the world. Being positioned on the subduction zone[20] with high seismic activity increases the possible threat of floods and landslides. Besides the country's fragile geographic setting, the unplanned settlements and poorly constructed buildings are additional risk factors that increase the country's vulnerability to the earthquakes.[21]

The Disaster Event Phase of the Disaster Process

On April 25, 2015, earthquake of magnitude 7.8 on the Richter scale struck the Kathmandu region in Nepal. It was the strongest earthquake to occur in Nepal since 1934.[22] The main impact was later followed with an aftershock on May 12, 2015. Both impacts lead to the death of 8891 people, 22,302 injured and left 375 people as

[19] Johnson and Olshansky (2016:32)

[20] Seizmical collision between the Indian and Eurasian tectonic plates

[21] Subedi and Chhetri (2019:2)

[22] AON (2015:6)

Table 6.2 Disaster process overview of the volcanic eruption of the earthquake and tsunami in Japan

<table>
<tr><th colspan="4">Disaster process overview</th></tr>
<tr><td colspan="2">Disaster category</td><td colspan="2">Natural</td></tr>
<tr><td colspan="2">Disaster sub-category</td><td colspan="2">Geophysical disasters: Earthquake and tsunami</td></tr>
<tr><td colspan="2">Case study disaster (country, region/city, date, year of disaster event)</td><td colspan="2">Japan, Tohoku region, March 11, 2011</td></tr>
<tr><td colspan="4">Disaster process overview—use of the urbanscape</td></tr>
<tr><td colspan="4">The pre-disaster phase
Prevention and mitigation</td></tr>
<tr><td colspan="2">Urban pattern of the everyday life</td><td colspan="2">Urban vulnerability</td></tr>
<tr><td colspan="2">Under continuous threat of earthquakes, making the Pacific coastline of the Tohoku region highly vulnerable to the impact of tsunamis
Development of resilient measures based on the past earthquake records: the construction of seawalls and flood gates; urban planning focused on the resilience; and evacuation planning</td><td colspan="2">The scale, density and construction of the existing urban development, by examining if the existing buildings, are constructed on built or natural earth, if the basements are occupied with parking spaces and if the density of the existing buildings limits the empty spaces within the urban structure</td></tr>
<tr><td colspan="4">The disaster event phase
Disaster impact—urbanscape identity disruption</td></tr>
<tr><td colspan="2">Direct disaster effect</td><td>Secondary effect</td><td>Indirect disaster effect</td></tr>
<tr><td>Population impact</td><td>A total of 15,879 deaths and 4086 missing persons</td><td rowspan="3">The earthquake induced a great tsunami along the Pacific side of the Tohoku region and floods, followed by a nuclear disaster in the Fukushima prefecture</td><td rowspan="3">The nuclear disaster in the Fukushima prefecture, lead to a long-term evacuation. Because of the level of radiation, many people were forced to leave their homes and 335,000 refugees were in need of essential supplies for survival</td></tr>
<tr><td>Property impact</td><td>The property damage was divided into the destruction of houses by seismic motion and the houses along the coast that got washed away
Six months after the earthquake, 125,530 buildings were destroyed and 240,332 damaged</td></tr>
<tr><td>Urban impact</td><td>Ruining 23,600 hectares of farmland and altering the coastline with the 0.6 m subsidence of 400 km of Japan's east coast
Damage to many road facilities, especially the main route along the coast, damages to bridge bearings and joints, the water drainage systems, electrical power station and electricity transmission facilities
Many bridge decks and road embankments got washed away</td></tr>
</table>

(continued)

Table 6.2 (continued)

The disaster event phase Disaster impact—urbanscape identity disruption			
Direct disaster effect		Secondary effect	Indirect disaster effect
Urban resilience	The open public spaces during the disaster impact enabled evacuation of people. Many green spaces were occupied with shelters. The urban pattern of the everyday life that enhances the movement within the existing roads influenced the evacuation routes during the disaster impact		
The post-disaster phase Post-disaster recovery—post-disaster urbanscape identity rehabilitation			
Rehabilitation program	Rehabilitation plan		Urban rehabilitation strategies
Collective relocation program Land readjustment program Public housing program Tsunami recovery zone program	Future land-use patterns to reduce the risk of vulnerability and risk for future tsunamis		Relocation of land use Raising the base elevation of land and the height of the flood bank Reconstruction of housing property Redevelopment of urban infrastructure and facilities
The post-disaster urban rehabilitation models			
Model of land-use relocation Model of land readjustment Model of urban reconstruction			

missing. About 605,254 houses were completely destroyed and 288,255 houses were partially damaged, in both impacts combined.[23] The earthquake induced 2780 landslides and ground cracks, causing immense damage to the agriculture land, forest and water resources.[24] About 2.2% of the forest land within the affected areas was lost. The landslides generated sediments loads and debris from the damaged buildings, causing widespread damage to the urban infrastructure and

[23] AON (2015:9)

[24] MoSTE (2015:6)

Fig. 6.2 Photo of the Miyagi Prefecture, before the disaster, during the disaster impact, in March 2011, and during the post-disaster recovery in 2014. (Authors, August 2022)

land degradation.[25] In the aftermath of the disaster, many people were displaced into relief camps that occupied the open green spaces in the Kathmandu valley, but even a month after the impacts, about 2, eight million people were in the need for shelter.[26] The mitigation process in the aftermath of the disaster was not well organized. Due to the poor construction and lack of road networks, the efforts undertaken for the rescue and safety of the population failed to reach people in time, especially the people affected in rural and mountain areas.[27]

The Post-disaster Phase of the Disaster Process

The post-disaster recovery phase was reflected through the reconstruction process with the aim to ensure environmentally responsible recovery and reconstruction as "Build Back Greener".[28] The program was based on several principles that included strategies required to build back a better and environmentally resilient country. They included the development of spatial planning in terms of reduction of disaster risks for the resettlement of the affected population and the relocation of agriculture and infrastructure of the affected areas, in the aftermath of the disaster; the reuse and recycle of the debris material; the management of the disposal of the non-usable debris material; the redesign and strategic reconstruction of the road network, especially in the rural areas; the improvement of water resources in case of water contamination; the integration of farms in forest land to restore the agriculture sector; and reforestation, by replanting the trees and vegetation in the disaster affected areas.[29] In response to the earthquake and in order to build resilience to future seismic impacts, the government introduced "The Open Cities Project". The project was based on mapping the road networks and boundaries, schools and health facilities, by using the open-source platform "Street Map". Over 130,000 buildings have been mapped based on their building type and construction characteristics, in order to achieve valuable data for future response and disaster risk reduction.

---

[25] MoSTE (2015:5)

[26] MoSTE (2015:1)

[27] Subedi and Chhetri (2019:7)

[28] MoSTE (2015:2)

[29] MoSTE (2015:13)

The Post-disaster Urban Rehabilitation Models

Based on the disaster process overview, the post-disaster rehabilitation of the earthquake in Nepal is reflected through several post-disaster urban rehabilitation models. The earthquake induced landslides that generated sediments loads and debris material and caused an immense damage to the agriculture land. In the post-disaster rehabilitation, it was necessary to integrate the agriculture farms into the forest, revealing the post-disaster rehabilitation model of relocating the agriculture land. The debris material that got generated into downstream courses caused environmental degradation and required management of its use and disposal, identifying the model of the reuse of the usable debris material and the model of the management of the non-usable debris material during disposal, in the post-disaster rehabilitation. Due to the poor construction of the existing road network, especially in the rural areas, it was necessary to reconstruct the existing and damaged roads, identifying the post-disaster rehabilitation model of the redesign of the road network. With the loss of forest land, it was necessary to replant the trees and vegetation, revealing the post-disaster rehabilitation model of reforestation. In the aftermath of the earthquake, 2.8 million people were in the need of shelter. In the post-disaster recovery, it was essential to develop spatial plans for their resettlement, revealing the post-disaster rehabilitation model of strategic population resettlement. In order to build resilience, after the earthquake, valuable data of the buildings and road networks within the affected region have been collected. By mapping the building types, their construction characteristics and road networks, the data would be used as a response and a reduction tool for future disaster impact, revealing the post-disaster rehabilitation model of urban resilience (Table 6.3; Fig. 6.3).

Table 6.3 Disaster process overview of the volcanic eruption of the earthquake and landslides in Nepal

Disaster process overview	
Disaster category	Natural
Disaster sub-category	Geophysical disasters: Earthquake and landslides
Case study disaster (country, region/city, date, year of disaster event)	Nepal, Kathmandu region, April 25, 2015

Disaster process overview—use of the urbanscape The pre-disaster phase Prevention and mitigation	
Urban pattern of the everyday life	Urban vulnerability
Nepal is considered a high disaster-prone country, with a significant record of 19 major earthquakes since the twelfth century The capital city Kathmandu is the most vulnerable city in the world	Positioned in a zone of high seismic activity increases the possible threat of floods and landslides and the unplanned settlements and poorly constructed buildings are additional risk factors that increase the country's vulnerability to the earthquakes

The disaster event phase Disaster impact—urbanscape identity disruption			
Direct disaster effect		Secondary effect	Indirect disaster effect
Population impact	8891 people died, 22,302 people were injured and 375 people were listed as missing, in both impacts	Earthquake aftershock on May 12, 2015 The earthquake induced 2780 landslides	The landslides generated sediment loads and debris from the damaged buildings, causing widespread damage to the urban infrastructure and land degradation In the aftermath of the disaster, many people were displaced, even a month after the impacts, about 2,eight million were in the need of shelter
Property impact	605,254 houses were completely destroyed and 288,255 houses were partially damaged, in both impacts		
Urban impact	Immense damage to the agriculture land, forest and water resources. About 2.2% of the forest land within the affected areas was lost		
Urban resilience	Due to the poor construction and lack of road networks, the efforts undertaken for the rescue and safety of the population failed to arrive on time, especially to the people affected in rural and mountain areas		

The post-disaster phase Post-disaster recovery—post-disaster urbanscape identity rehabilitation		
Rehabilitation program	Rehabilitation plan	Urban rehabilitation strategies
"Build Back greener" "The open cities project"	Strategies required to build back a better and environmentally resilient country Mapping the road networks and	The development of spatial planning in terms of reduction of disaster risks for the resettlement of the affected population and the relocation of

(continued)

Table 6.3 (continued)

The post-disaster phase Post-disaster recovery—post-disaster urbanscape identity rehabilitation		
Rehabilitation program	Rehabilitation plan	Urban rehabilitation strategies
	boundaries, schools and health facilities, by using the open-source platform "street map"	agriculture and infrastructure of the affected areas, in the aftermath of the disaster The reuse and recycle of the debris material The management for the disposal of the non-usable debris material The redesign and strategic reconstruction of the road network, especially in the rural areas The improvement of water resources in case of water contamination Integrating farms in forest land to restore the agriculture sector; and reforestation, by replanting the trees and vegetation in the disaster affected areas Over 130,000 buildings have been mapped based on their building type and construction characteristics, in order to achieve valuable data for future response and disaster risk reduction
The post-disaster urban rehabilitation models		
Model of relocation of agriculture land Model of reuse of debris material Model of the disposal of the non-usable debris material Model of redesign of road networks Model of reforestation Model of strategic population resettlement Model of urban resilience		

Fig. 6.3 Photo of the Dharahara tower square in Kathmandu, before the disaster on October 25, 2014; during the disaster impact on April 27, 2015; and open public space in the Kathmandu valley, being occupied by one of the relief camps during the post-disaster recovery in 2015. (Authors, August 2022)

References

AON (2015) 2015 Nepal earthquake event recap report. Aon Benfield impact forecasting. AON, London

Eleazar L (2010) Philippines in land tenure and natural disasters. In: Addressing land tenure in countries prone to natural disasters. FAO, Rome, pp 75–96

Gaillard J, Le Masson V (2007) Traditional societies' response to volcanic hazards in the Philippines. Mt Res Dev 27(4):313–317. https://doi.org/10.1659/mrd.0949

Guzman EM (2004) Eruption of Mount Pinatubo in the Philippines in June 1991. ADRC, Kobe

Johnson LA, Olshansky RB (2016) After great disasters: how six countries managed community recovery. Lincoln Institute of Land Policy, Cambridge

Kazama M, Noda T (2012) Damage statistics (summary of the 2011 off the Pacific coast of Tohoku earthquake damage). Soils Found 52(5):780–792. https://doi.org/10.1016/j.sandf.2012.11.003

Koshimura S, Nobuo S (2015) Response to the 2011 Great East Japan earthquake and tsunami disaster. Philos Trans R Soc A Math Phys Eng Sci 373(2053):1–15. https://doi.org/10.1098/rsta.2014.0373

MoSTE (2015) Nepal earthquake 2015. Rapid environmental assessment. Ministry of Science, Technology and Environment, MoSTE, New Delhi

Norio O, Ye T, Kajitani Y, Shi P, Tatano H (2011) The 2011 eastern Japan great earthquake disaster: overview and comments. Int J Disaster Risk Reduct 2:34–42. https://doi.org/10.1007/s13753-011-0004-9

OECD (2006) OECD studies in risk management, Japan earthquakes. OECD, Paris

Omićević N, Bojanić Obad Šćitaroci B (2018) The urbanscape duality during a disaster and post-disaster. Prostor 26:106–115. https://doi.org/10.31522/p.26.1(55).8

Saxena M (2016) Role of open spaces in disaster management. AGORA, Baltimore

Subedi S, Chhetri MBP (2019) Impacts of the 2015 Gorkha earthquake: lessons learnt from Nepal. In: Earthquakes-impact, community vulnerability and resilience. IntechOpen, London. https://doi.org/10.5772/intechopen.85322

Chapter 7
Comparative Analysis of the Selected Climatological Disaster Process Case Studies

Abstract This chapter presents an analytical overview of the selected climatological disaster case studies in the world that include droughts, extreme temperatures and wildfires. Within the climatological disaster natural subcategory, the following post-disaster case studies are analysed: Wildfires in Indonesia (1997); Wildfires in California, the United States of America (2017; 2018) and Heat Wave and Wildfires in Russia (2010), based on the use of the urbanscape, through its built and natural elements: the urban pattern of the everyday life and the urban vulnerability during the pre-disaster phase; the direct urban impact during the disaster event phase and the urban rehabilitation during the post-disaster phase.

Keywords Natural disasters · Climatological disasters · Wildfires · Heat wave

7.1 Comparative Analysis of the Selected Climatological Disaster Process Case Studies

7.1.1 Wildfires in Indonesia (1997)

The Pre-disaster Phase of the Disaster Process

Indonesia is the country most affected by wildfire in Southeast Asia, since the climate is strongly under the influence of the El Nino Southern Oscillation (ENSO).[1] The El Nino circulation effects cause droughts in Southeast Asia, leading to uncontrolled fires.[2] During the dry season, most of the fires in Indonesia occur in Kalimantan and Sumatra, since they contain the world's largest areas of tropical peatland.[3] The fire severity increased over history, mainly due to the uncontrolled burning of land, reflected through the cleaning of forest land through their

[1] The ENSO refers to the period variations in wind and sea surface temperature over the Pacific Ocean, whereby the warming phase is known as the El Nino and the cooling phase as the La Nina.

[2] Hirschberger (2017)

[3] Harrison et al. (2009:156)

N. Omićević, B. Bojanić Obad Šćitaroci, *The Urban Rehabilitation of Post-Disaster Scapes*, Disaster Resilience and Green Growth,
https://doi.org/10.1007/978-981-19-9505-7_7

conversion into agricultural land. Since Indonesia is the world's largest palm-oil producer, the country's deforestation is being used as the tool for producing palm-oil plantations.[4] Indonesia covers 22.5 million of peatland territory.[5] Peatlands are types of wetlands that are considered to be the most valuable ecosystems on Earth. They preserve the global biodiversity by storing more carbon than all other vegetation types in the word, ensure safe drinking water and decrease the risk of floods, but their damage as the result of agricultural conversion and burning contributes to the emission of carbon dioxide (CO_2), accelerating global warming.[6] The fires in 1997 alone caused the emission of 2.2 billion tons of CO_2.[7] This resulted from the lack of awareness for the protection and restoration of peatlands.

The Disaster Event Phase of the Disaster Process

In 1997, Indonesia experienced the most severe wildfires worldwide. Due to the fire's carbon emission of the affected peat lands, the disaster caused a thick haze over Southeast Asia and led to one of the worst environmental disasters of the century. For several months, the smoke covered Indonesia, Malaysia, Singapore, Brunei, southern part of Thailand and the Philippines.[8] The air pollution caused short-term and long-term health effects of the population and led to over 15,600 child, infant and fetal deaths.[9] During the fire, up to ten million hectares of forest and non-forest land was affected in Indonesia. Not all forest fires have a negative impact on the forest ecosystem. They can lead to the formation of different types of savannas and grasslands and regenerate into post-fire secondary forests.[10]

The Post-disaster Phase of the Disaster Process

In the aftermath of the fire, a large scale of unplanned forest degradation took place that decreased the process of the forest recovery. The post-disaster rehabilitation program was focused on pursuing the natural recovery of the tropical forest ecosystem. This implied rehabilitation methods that include the availability of seeds for the development of the post-fire secondary forest, the importance of natural regeneration through mixed planting of native species and the awareness of the environmental importance of the secondary forests.[11] During the post-disaster phase, at least 2.3

[4] Harrison et al. (2009:157)
[5] Hirschberger (2017:84)
[6] IUCN (2021)
[7] Hirschberger (2017:88)
[8] Sastry (2002)
[9] Jayachandran (2008)
[10] Dennis et al. (2001:788)
[11] Dennis et al. (2001:792)

million ha of potential post-fire secondary forest land appeared in East Kalimantan. If recovered, this post-fire secondary forest land would become an important resource in terms of providing timber and non-timber products.[12] In order to achieve the natural recovery and prevent future forest fires, it was important to improve the organization of the land use and plans for small case farms, forestry and plantations, cultivating the land without fire and recovering the natural groundwater in the peat dams in order to reduce the fire vulnerability.[13]

The Post-disaster Urban Rehabilitation Models

Based on the disaster process overview, the post-disaster rehabilitation of the wildfires in Indonesia is reflected through several post-disaster urban rehabilitation models. Indonesia contains the world's largest areas of tropical peatlands, as the most valuable ecosystem on Earth, covering 22.5 million of peatland territory. Due to the country's agricultural land conversion and burning, the damage of peatlands leads to the emission of a considerable amount of carbon dioxide, revealing the post-disaster urban rehabilitation model of the lack of awareness for the preservation of the natural tropical ecosystem. Being the world's largest palm-oil producer, the country's forest land is cleaned and converted into palm-oil plantations, identifying the post-disaster urban rehabilitation model of deforestation and land conversion. Because of the uncontrolled burning of forest and non-forest land through history, the country has been exposed to an increasing impact of wildfires, by revealing the post-disaster urban rehabilitation model of continuity. Despite their increased severity, during the post-disaster recovery, wildfires can influence the formation of different types of savannas and grasslands through the development of post-fire secondary forests, enriched with planting of native species, revealing the post-disaster urban rehabilitation model of resilience through natural regeneration. The recovery of the post-disaster secondary forest in the post-disaster rehabilitation would contribute to the environmental awareness of its importance and become a valuable resource in the future use of forest land, revealing the post-disaster urban rehabilitation model of natural recovery (Table 7.1; Fig. 7.1).

---

[12] Dennis et al. (2001:793)

[13] Hirschberger (2017:90)

Table 7.1 Disaster process overview of the wildfires in Indonesia

Disaster process overview	
Disaster category	Natural
Disaster sub-category	Climatologic disasters: wildfire
Case study disaster (country, region/city, date, year of disaster event)	Indonesia, 1997

Disaster process overview—use of the urbanscape	
The pre-disaster phase Prevention and mitigation	
Urban pattern of the everyday life	Urban vulnerability
Indonesia is the most affected country of wildfire in Southeast Asia, since the climate is strongly under the influence of the El Nino Southern Oscillation that causes droughts in Southeast Asia, leading to uncontrolled fires Most of the fires in Indonesia occur in Kalimantan and Sumatra, since they contain the world largest areas of tropical peatland. Their severity increased during history, mainly due to the uncontrolled burning of land, reflected through the cleaning of forest land through their conversion into agricultural land	The country's deforestation is being used as the tool for producing palm-oil plantations The lack of awareness for the protection and restoration of peatlands as the most valuable ecosystems on Earth They preserve the global biodiversity by storing more carbon than all other vegetation types in the word, ensure safe drinking water and decrease the risk of floods, but their damage as the result of agricultural conversion and burning contributes to the emission of carbon dioxide, (CO_2)

The disaster event phase Disaster impact—urbanscape identity disruption			
Direct disaster effect		Secondary effect	Indirect disaster effect
Population impact	Over 15,600 child, infant and fetal deaths	The fire's carbon emission of the affected peatland caused a thick haze over Southeast Asia	One of the worst environmental disasters of the century The fires caused the emission of 2.2 billion tons of CO_2 The air pollution caused short-term and long-term health effects
Property impact	/		
Urban impact	During the fire, up to ten million ha of forest and non-forest land was affected in Indonesia		
Urban resilience	The forest fires can lead to the formation of different types of savannas and grasslands and regenerate into post-fire secondary forests		

The post-disaster phase Post-disaster recovery—post-disaster urbanscape identity rehabilitation		
Rehabilitation program	Rehabilitation plan	Urban rehabilitation strategies
The post-disaster rehabilitation program was focused on pursuing the natural recovery of the tropical forest ecosystem	Recovery of the post-fire secondary forest land	The availability of seeds for the development of the post-fire secondary forest The importance of natural

(continued)

Table 7.1 (continued)

The post-disaster phase Post-disaster recovery—post-disaster urbanscape identity rehabilitation		
Rehabilitation program	Rehabilitation plan	Urban rehabilitation strategies
		regeneration through mixed planting of native species and the awareness of the environmental importance of the secondary forests
The post-disaster urban rehabilitation models		
Model of lack of awareness for the preservation of peatlands Model of deforestation and land conversion Model of continuity Model of resilience by post-fire natural regeneration Model of natural recovery		

7.1.2 *Wildfires in California, The United States of America (2017; 2018)*

The Pre-disaster Phase of the Disaster Process

Unlike other natural disasters, wildfires are perceived to be prevented. With the limited and controlled burning, strategic locations are built to defend the fire breaks and to mitigate the fire. These fire breaks are the open spaces created in the forest that enabled the continuity of fuels in a controlled manner. This reduces the fire impact and at the same promotes diversity in the reestablishment of the existing vegetation.[14] Out of many California's wildfires, the Thomas wildfire that evoke in December 2017 was one of the largest and most destructive in the history of California. The main impact factors that contributed to the disaster risk of the California wildfires consist of the growth of the residential areas; the Santa Ana and Diablo Winds[15]; the lack of controlled fire burning; the steep topography; vegetation spacing and presence of flammable vegetation.[16]

The Disaster Event Phase of the Disaster Process

The fire burned 281,900 acres of land, out of which 181,300 acres were National Forest lands; 98,200 acres were private lands and 2400 acres were a combination of state and country properties. The extensive property impact on approximately 1063

---

[14] Saxena (2016)

[15] Diablo Winds are hot and dry winds that occur in North California, during Spring and Fall. Santa Ana Winds are strong and extremely dry winds in South California

[16] AGCS (2018)

Fig. 7.1 Photo of aerial view of the deforestation of nearly 41,440 km^2 rainforest land in Kalimantan, since 1973, during the pre-disaster phase; the fire in the Indonesian peatlands during the disaster in 1997; and the satellite image of the air haze pollution across the Indian Ocean, in the aftermath of the disaster, on October 22, 1997. (Authors, August 2022)

structures led to evacuation of thousands of people and displacement of animals that influenced the small rate of population impact. The highest impact was on the natural landscape. Since more than half of the burned area had not experienced fire within the last 30 years, the existing dense and thick vegetation increased the disaster impact even more. The fire burned the supporting vegetation of the rocks that protect the slope, changed the soil structure and compromised the stability of the landscape. This landscape alteration increased the post-disaster vulnerability with the appearance of debris flows and debris slides. Consequently, all post-fire hazards can affect human lives, the habitat and water quality, and cause damages to the urban infrastructure and residential properties within the area.[17]

The Post-disaster Phase of the Disaster Process

After the Thomas Fire, the wildfires in California evoked again in November 2018 as the Camp Fire occupied the Butte County and the Woolsey Fire occupied the Ventura County. The Camp Fire caused burning of 153,336 acres of land, destruction of 18,804 structures and the death of 86 people. The Woolsey Fire caused burning of 96,949 acres of land, destruction of 1643 structures and the death of three people.[18] According to official numbers, the 2018 California wildfires caused 100 deaths.[19] For this reason, the post-disaster rehabilitation is still in progress. The post-disaster rehabilitation strategy consists of the development of a wildfire mitigation plan before the new fire season occurs. This plan includes the assessment of the contribution factors for the wildfires and the analysis of the "Burned Area Emergency response" team (BAER).[20] The BAER team report consisted of two analytical phases: the assessment of the burned area and the establishment of the Landslide Hazards Program, after the assessment. The Landslide Hazards Program conducted a debris flow assessment to obtain the probability model of evoking debris flows as the most dangerous post-fire hazard and the volume model to estimate its volume in occurrence of rainfall.

The Post-disaster Urban Rehabilitation Models

Based on the disaster process overview, the post-disaster rehabilitation of the wildfires in California is reflected through several post-disaster urban rehabilitation models. Due to the Santa Ana and Diablo Winds and the lack of controlled fire burning, the state of California has been exposed to a historical record of wildfires, revealing the post-disaster urban rehabilitation model of continuity. With limited and

[17] King and Veich (2018)

[18] www.fire.ca.gov

[19] https://www.fire.ca.gov/incidents/2018/

[20] King and Veich (2018:1)

controlled burning during the pre-disaster phase, strategic locations are built to reduce the fire impact during the disaster impact, identifying the post-disaster urban rehabilitation model of resilience. During the post-disaster recovery, before the new fire season starts, it was necessary to develop a wildfire mitigation plan that would show the main disaster risk factors, revealing the post-disaster urban rehabilitation model of the vulnerability assessment. The vulnerability assessment was based on the analytical approach of examining the burnt area and the probability of evoking debris flow, revealing the post-disaster urban rehabilitation models of the burnt area assessment and the post-disaster urban rehabilitation model of estimating debris flows (Table 7.2; Fig. 7.2).

7.1.3 Heat Wave and Wildfires in Russia (2010)

The Pre-disaster Phase of the Disaster Process

Russia as the largest country in the world covers the largest forest land, 22% of the world forests[21] and peatland, occupying 8% of the country's territory. Every year several million hectares of forest land is burned leading to deforestation and several million hectares of peatland in the European part of the country is drained and used for agriculture, forestry and peat extraction.[22] The annual repeated fires increased the deforested land by eight million hectares in the past 50 years.[23] The drained peatlands oxidize and release a large amount of carbon dioxide (CO_2) into the atmosphere, resulting with a negative impact on the climate and biodiversity.

The Disaster Event Phase of the Disaster Process

In 2010, Moscow and the central part of Russia were under the impact of an unprecedented heat wave with extreme drought that combined with strong winds led to the outbreak of several wildfires between June 21 and August 19.[24] With the temperature of 45 °C throughout the month of July, it was the worst drought recorded in the last 130 years.[25] The fires devastated the forest and crop land. About 30,376 forest fires were registered, including 1162 peat fires.[26] About five million hectares of forest land was burned in 2010,[27] 3180 houses were destroyed

[21] Bockel et al. (2014:5)
[22] ICI (2013:1)
[23] Hirschberger (2017:67)
[24] European Union (2011:58)
[25] Mokhnacheva (2011:29)
[26] Munich RE (2015:1)
[27] Hirschberger (2017:65)

Table 7.2 Disaster process overview of the wildfires in California

Disaster process overview			
Disaster category		Natural	
Disaster subcategory		Climatologic disasters: wildfire	
Case study disaster (country, region/city, date, year of disaster event)		United States of America, Pacific region/California, December, 2017; November, 2018	
Disaster process overview—use of the urbanscape			
The pre-disaster phase Prevention and mitigation			
Urban pattern of the everyday life		Urban vulnerability	
Out of many California's wildfires, the Thomas wildfire that evoke in December 2017 was one of the largest and most destructive in the history of California		The main impact factors that contributed to the disaster risk of the California wildfires consist of the growth of the residential areas; the Santa Ana and Diablo Winds; lack of controlled fire burning; the steep topography; vegetation spacing and presence of flammable vegetation Since more than half of the burned area did not experience fire within the last 30 years, the existing dense and thick vegetation increased the disaster impact even more	
The disaster event phase Disaster impact—urbanscape identity disruption			
Direct disaster effect		Secondary effect	Indirect disaster effect
Population impact	100	Landscape alteration that increased the appearance of post-fire hazards such as debris flows and debris slides	Evacuation and displacement of people and animals Disabling the function of the urban infrastructure with the appearance of debris flows and debris slides that could damage the transport and communication facilities within this vulnerable area
Property impact	The fire burned 281,900 acres of land, out of which 181,300 acres were National Forest lands; 98,200 acres were private lands; 2400 acres were a combination of state and country properties that included an extensive property impact of approximately 1063 structures		
Urban impact	The fire burned the supporting vegetation of the rocks that protect the slope, changed the soil structure and compromised the stability of the landscape The infrastructure impact can be caused indirectly, since the landscape alteration can increase the post-disaster vulnerability with the appearance		

(continued)

Table 7.2 (continued)

The disaster event phase Disaster impact—urbanscape identity disruption			
Direct disaster effect		Secondary effect	Indirect disaster effect
	of debris flows and debris slides that could damage the transport and communication facilities within this vulnerable area		
Urban resilience	With the limited and controlled burning, strategic locations are built to defend the fire breaks and to mitigate the fire These fire breaks are the open spaces created in the forest that enable the continuity of fuels in a controlled manner that reduces the fire impact, and at the same promotes diversity in the reestablishment of the existing vegetation		

The post-disaster phase Post-disaster recovery—post-disaster urbanscape identity rehabilitation		
Rehabilitation program	Rehabilitation plan	Urban rehabilitation strategies
Burned Area Emergency Report Landslide Hazards Program	Fire mitigation plan to prevent and reduce the risk of fire before the new season starts, that is still in progress	Assessment of the burned area Examining the possibility of evoking post-fire hazards, such as debris flows, based on the probability model and the volume model

The post-disaster urban rehabilitation models
Model of urban resilience through controlled burning Model of continuity Model of vulnerability assessment Model of assessment of the burned area Model of estimating debris flows

and 62 people killed.[28] Thousands of people were forced to seek temporary shelters which resulted in the resettlement of nearly 100,000 people from the rural parts of Russia.[29] The toxic smoke resulting from the peat fires released 140 million tons of

[28] Mokhnacheva (2011:30)
[29] Mokhnacheva (2011:27)

Fig. 7.2 Photo of aerial view of the Coffey Park neighbourhood in Santa Rosa, California, before the disaster impact; during the disaster impact in December, 2017; and the aerial view of the completely destroyed neighbourhood in Paradise, California, in November 2018, during the post-disaster rehabilitation progress. (Authors, August 2022)

carbon dioxide (CO_2)[30] that led to a significant health impact and increased the mortality rate. The country responded to the disaster with the necessary infrastructure that included over 30,000 trucks, 200 aircrafts and the help of 200,000 fire fighters to mitigate the fast-spreading wildfires.[31]

The Post-disaster Phase of the Disaster Process

The post-fire recovery process was reflected through several programs and projects. The development of the Forest Fire Response Project[32] was established to improve the prevention and to enhance the management of forest fire in order to raise the awareness of its control and mitigation. The rehabilitation strategies included the establishment of protected areas within the forest and replanting larches, since they are resistant and provide no fuel for smaller fires.[33] In addition to the natural replanting rehabilitation strategies, the Russian Government developed a large-scale rewetting program of drained peatlands in the central region of Russia. The peatland rehabilitation implied the regeneration of the natural vegetation to restore their natural ecosystem. The rewetting program included the restoration of more than 40,000 hectares of degraded peatland and the development of guidelines for the peatland conservation and rehabilitation.[34]

The Post-disaster Urban Rehabilitation Models

The disaster process overview of the heatwave and wildfires in Russia reveals several post-disaster urban rehabilitation models. Due to the increased deforestation and drainage of peatlands, Russia is losing several millions of hectares of forest and peatland every year, revealing the post-disaster urban rehabilitation model of continuity through the continuous wildfire impact. In order to improve the prevention and enhance the management of forest fire, it was essential to raise the awareness of its control and mitigation, identifying the post-disaster urban rehabilitation model of raising awareness for fire mitigation. The rehabilitation strategies in the post-disaster recovery were mainly focused on the enhancement of the forest and peatland recovery, revealing the post-disaster urban rehabilitation model of natural recovery, through the establishment of protected areas within the forest and replanting larches, identifying the post-disaster urban rehabilitation model of reforestation and replanting. The rewetting program of the drained peatland in the post-disaster rehabilitation phase implied the regeneration of the natural vegetation in order to restore their natural ecosystem, revealing the post-disaster urban rehabilitation model of peatland restoration and conservation (Table 7.3; Fig. 7.3).

---

[30] Hirschberger (2017:68)

[31] European Union (2011:58)

[32] Bockel et al. (2014:5)

[33] Hirschberger (2017:69)

[34] ICI (2013:2)

Table 7.3 Disaster process overview of the heat wave and wildfires in Russia

Disaster process overview	
Disaster category	Natural
Disaster subcategory	Climatologic disasters: heat wave and wildfire
Case study disaster (country, region/city, date, year of disaster event)	Russia, June 21–August 19, 2010

Disaster process overview—use of the urbanscape

The pre-disaster phase
Prevention and mitigation

Urban pattern of the everyday life	Urban vulnerability
Russia covers the largest forest land, 22% out of the world forests and peatland occupying 8% of the country's territory. Every year several million ha of forest land gets burned leading to deforestation and several million ha of peatland in the European part of the country is being drained and used for agriculture, forestry and peat extraction	The annual repeated fires increased the deforested land by eight million ha in the past 50 years The drained peatlands oxidize and release a large amount of carbon dioxide, (CO_2) into the atmosphere, resulting in a negative impact on the climate and biodiversity

The disaster event phase
Disaster impact—urbanscape identity disruption

Direct disaster effect		Secondary effect	Indirect disaster effect
Population impact	62 people died	The fires devastated the forest and crop land The toxic smoke resulting from the peat fires released 140 million of carbon dioxide	The resettlement of nearly 100,000 people from the rural parts of Russia
Property impact	3180 houses were destroyed		
Urban impact	The fires devastated the forest and crop land. About 30,376 forest fires were registered, including 1162 peat fires About five million ha of forest land was burned		
Urban resilience	The country responded to the disaster with the necessary infrastructure that included over 30,000 trucks, 200 aircrafts and the help of 200,000 fire fighters to mitigate the fast-spreading wildfires		

The post-disaster phase
Post-disaster recovery—post-disaster urbanscape identity rehabilitation

Rehabilitation program	Rehabilitation plan	Urban rehabilitation strategies
The Forest Fire Response Project The large-scale	To improve the prevention and to enhance the management of forest fire in order to raise the awareness	The establishment of protected areas within the forest and replanting larches

(continued)

Table 7.3 (continued)

The post-disaster phase		
Post-disaster recovery—post-disaster urbanscape identity rehabilitation		
Rehabilitation program	Rehabilitation plan	Urban rehabilitation strategies
rewetting program of drained peatlands	of its control and mitigation The regeneration of the natural vegetation to restore their natural ecosystem	The restoration of more than 40,00 ha of degraded peatland and the development of guidelines for the peatland conservation and rehabilitation
The post-disaster urban rehabilitation models		
Model of continuity Model of raising awareness for wildfire mitigation Model of natural recovery Model of reforestation and replanting Model of peatland restoration and conservation		

Fig. 7.3 Photo of aerial view of the wild taiga forest in northern Russia, during the pre-disaster phase; during the disaster impact in August 2010; and the rewetting of the Ostrovskoye peatland site in the Sobinsky District, Vladimir Region in March 2019 during the post-disaster rehabilitation progress. (Authors, August 2022)

References

AGCS (2018) Burning issues. California Wildfire review. Allianz Global Corporate & Speciality SE, Paris

Bockel L, Bernoux M, Thapa D, Armstrong A (2014) GHG and natural capital impact of Russia forest fire response project. FAO, Rome

Dennis RA, Hoffmann A, Applegate G, von Gemmingen G, Kartawinata K (2001) Large-scale fire: creator and destroyer of secondary forests in Western Indonesia. J Trop For Sci 13(4):786–799

European Union (2011) Forest Fires in Europe 2010. Institute for Environment and Sustainability Land Management and Natural Hazards Unit. Publications Office of the European Union, Luxembourg

Harrison M, Page S, Limin S (2009) The global impact of Indonesian forest fires. Biologist 56(3): 156–163

Hirschberger P (2017) Forests ablaze, causes and effects of global forest fire. The World Wide Fund for Nature, Berlin

https://www.fire.ca.gov/incidents/2018/. Accessed 20 Aug 2022

https://www.fire.ca.gov/incidents/2018/11/8/woolsey-fire/. Accessed 17 Aug 2022

ICI (2013) Restoring peatlands in Russia. Preventing peat fires, mitigating climate change. Wetlands International, Wageningen

IUCN (2021) Peatlands and climate change. Issues brief. IUCN, Gland

Jayachandran S (2008) Air quality and early-life mortality: evidence from Indonesia's wildfires. J Hum Resour 44(4):916–954. https://doi.org/10.3368/jhr.44.4.916

King A, Veich D (2018) Thomas Fire. Los Padres National Forest. Burned Area Emergency Response (Baer) Assessment Specialist Report, Phase 2—Geologic hazards. United States Department of Agriculture, Washington, DC

Mokhnacheva D (2011) Wildfires in Russia. In: The state of environmental migration 2010. IDDRI, Paris, pp 27–38

Munich RE (2015) Heat wave, drought, wildfires in Russia (Summer 2010). Geo-risk research. Munich RE, Munich

Sastry N (2002) Forest fires, air pollution, and mortality in Southeast Asia. Demography 39(1): 1–23. https://doi.org/10.2307/3088361

Saxena M (2016) Role of open spaces in disaster management. AGORA, Baltimore

Chapter 8
Comparative Analysis of the Selected Sociological Man-Made Disaster Process Case Studies

Abstract This chapter presents an analytical overview of the selected sociological man-made disaster case studies in the world, caused by intentional human actions, that include terrorism, conflicts of war and genocide. Within the sociological man-made disaster subcategory, the following disaster case studies are analysed: The Siege of Sarajevo (1992–1996); The Battle of Grozny (1994–1995) and The September 11 Terrorist Attacks in New York (2001), based on the use of the urbanscape, through its built and natural elements: the urban pattern of the everyday life and the urban vulnerability during the pre-disaster phase; the direct urban impact during the disaster event phase and the urban rehabilitation during the post-disaster phase.

Keywords Man-made disasters · Sociological disasters · War · Terrorism

8.1 Comparative Analysis of the Selected Sociological Man-Made Disaster Process Case Studies

8.1.1 The Siege of Sarajevo (1992–1996)

The Pre-disaster Phase of the Disaster Process

The pre-disaster phase was mainly reflected through the peaceful urban environment provided through the safety and protection within the urbanscape. During this period, the national defence and protection of possible impact of disasters was discussed in the "Program of Construction and Urban Development of Sarajevo for the Period 1971–1985". This study was based on defining the factors for survival and evacuation during the disaster impact, in terms of identifying the appropriate roads for evacuation and construction of facilities that would provide rescue and shelter needs.[1] In these terms, the longitudinal urban pattern of Sarajevo's spatial

[1] Pilav (2012:24)

N. Omićević, B. Bojanić Obad Šćitaroci, *The Urban Rehabilitation of Post-Disaster Scapes*, Disaster Resilience and Green Growth,
https://doi.org/10.1007/978-981-19-9505-7_8

development represented a disadvantage in case of evacuation. Since airports are one of the main impact targets during any war, the location of Sarajevo's airport was also assessed. In case of war, the airport's location would represent a danger for its citizens since it is located in a densely populated urban environment within the city. The recommendations for coping with the disaster were based on creating a plan of higher and lower zones of risk within the city; implementing self-supply stations in case of the need for water or electricity supply, with each zone accessible to two supply points; establishing shelters for people that would have to remain in the city during the disaster, by providing one shelter space for each citizen at home or at work and establishing shelters in the centre of the city, such as pedestrian tunnels, traffic roads, underground garages and buildings with dualistic purposes that would be used in the everyday life in the pre-disaster phase and during the disaster impact.[2] Since this study remained undeveloped, the lack of existing pre-disaster risk reduction plans and the necessary strategies and mitigation tools to cope with the disaster events, increased the vulnerability for the disaster impact even more.

The Disaster Event Phase of the Disaster Process

The Siege of Sarajevo as one of the longest in modern history, lasted 4 years and had a devastating impact on the city. During the Siege, 10,514 civilians of which 1598 were children, died or disappeared. In addition, 1741 people of which 356 were children, were injured or became permanently disabled.[3] The Serb forces continuously attacked the city from the occupied hills with an average of 329 shells per day. This shell impact caused an extensive destruction to the city's physical properties, destroying over 10,000 apartments and damaging 100,000 other buildings. According to the United Nations Report of the Study of the Battle and Siege of Sarajevo, 23% of the property destruction was reported as seriously damaged, 64% as partially damaged and 10% as slightly damaged.[4] During the Siege and after the War, many green spaces and parks were transformed into cemeteries and others have been deliberately contaminated by the aggressor with landmines. Due to the lack of heating supply, three-quarters of all urban trees and nearly all peri-urban trees within the siege line were cut down for firewood and household needs.[5] In terms of the urban infrastructure impact, during the Siege many buildings were transformed into temporary shelters, such as schools that were forced to hold classes in apartments or building hallways. The main tram lines, buses and trolleys in the city were not able to function, only the Sarajevo airport, controlled by the United Nations Protection Force and the underground corridor underneath the airport, served as the main public transportation. This underground corridor was used mainly for military transport,

[2] Pilav (2012:25)
[3] Ernst et al. (2003)
[4] William (1994)
[5] Lacan and McBride (2009)

essential supplies for survival and transport of the injured.[6] The main tool used for the disaster impact was in the form of cutting of phone lines, electricity, water and gas. Due to the cut-offs of drinking water and electricity, the citizens of Sarajevo were forced to find other sources of water supplies. The river Miljacka was used for washing clothes, and the waiting lines for getting water from alternative water sources were often under the impact of the surrounding snipers positioned on the hillsides. In terms of the social impact, about half of 4.4 million Bosnian populations were displaced during the Bosnian War (1992–1995). Because of this external and internal displacement, the urban development of Sarajevo had to meet the required need for housing.

During the Siege, the surrounding hillside of Sarajevo's urbanscape was used as a strategic position in terms of the impact from the aggressor and at the same time as a defensive line of resilience. Situated in the valley, the city favoured the defensive strategy in terms of creating a belt of urban resilience but at the same time it enabled a more intense impact from the aggressor's side. The urban pattern of the everyday life shifted to another layer, below the city, in form of an underground corridor. The familiar movement within the existing urban structure influenced the survival routes, for water and food supply. For this reason, they were used as resilience routes, transformed to misguide the aggressor form the surrounding hillside, since they were continuously under the shell impact.[7] Apart of destroying nearly all buildings in the city, the Serb forces deliberately attacked places of historical and religious importance, the destruction of which would cause the destruction of a collective identity. One of these deliberate identity impacts was the destruction of the Bosnian National and University Library, that resulted with the loss of 700 manuscripts and a unique collection of Bosnian publications with the main aim to destroy the evidence of history.[8]

The Post-disaster Phase of the Disaster Process

The various aspects of the post-disaster rehabilitation process of Sarajevo are reflected in the post-disaster reconstruction that is still in progress; memorialization of the War through monuments and memorials; re-redefining the urbanscape identity; and the possibility of reconciliation with the past. The post-disaster reconstruction consists of two plans: the Spatial Plan of Canton Sarajevo (1992–2015) and the Spatial Plan of Canton Sarajevo (2003–2023). It is based on the reconstruction of the urban infrastructure; housing reconstruction, reflected through the post-war informal constructions that increased future environmental hazards and relocation of population; reconstruction of public buildings; and the reconstruction of religious buildings and buildings with cultural and historical values, reflected through the post-war

---

[6] Pilav (2012)

[7] Omićević and Bojanić Obad Šćitaroci (2018)

[8] Troncota (2015)

division of monuments and heritage.[9] The re-defined urbanscape identity evolved from the post-disaster spatial alteration, that occurred through the construction of massive and dense multi-storey apartment blocks in the western part of the city; the urban sprawl on the surrounding slopes; alteration of the traditional landscapes[10]; and the destruction of open public and green space and their transformation into cemeteries. During its post-war progress (1995-present), Sarajevo's urbanization developed according to the large population growth. Due to the large demographic inflow following the war, this urban process was manifested through the urban and suburban sprawl on the city's surrounding slopes. The spatial distribution of the population growth was under the impact of natural and social forces and that strongly affected the population to settle on the city's slopes.[11] The new urban population urbanized the slopes through the expansion of informal settlements. The urban sprawl of post-war Sarajevo involved the construction of 25,000–50,000 informal residential units.[12]

The Post-disaster Urban Rehabilitation Models

Based on the disaster process overview, the post-disaster rehabilitation of the Siege of Sarajevo is reflected through several post-disaster urban rehabilitation models. Although the study for the protection of possible disaster impact was discussed in the pre-disaster phase, it was never carried out. The discussed mitigation strategies related to the disadvantage of the city's longitudinal urban pattern and the airport being positioned in a densely populated urban environment, providing shelters and buildings with dualistic purposes to be used in the pre-disaster phase and during the disaster impact, proved to be extremely important in terms of both, the urban impact and urban resilience. The lack of existing pre-disaster risk reduction and the necessary strategies to cope with possible disaster events increased the city's vulnerability to the disaster impact even more, which reveals the post-disaster urban rehabilitation model of raising the awareness for the disaster risk reduction and mitigation. The dualistic role of Sarajevo's urbanscape and the urban pattern of the everyday life influenced the evacuation routes, the development of the defensive line on the surrounding hillsides and the belt of urban resilience in the valley, but it also enabled the strategic position of the aggressor on the surrounding hillsides to have a more intense impact, with the city being situated in the valley and the urban pattern of the everyday life shifting to the underground, below the city, which identifies the post-disaster urban rehabilitation model of urbanscape dualism in terms of urban resilience and urban impact. During the post-disaster rehabilitation progress, the city's

---

[9]Englund (2015)

[10]Traditional neighbourhood units—the *mahalas*, that were built on the slopes of Sarajevo during the Ottoman period, (1463–1878)

[11]Nurković (2015:26)

[12]Martín-Díaz et al. (2015:380)

urbanscape was transformed through several spatial alterations, which re-defined its urbanscape identity that required the post-disaster urban rehabilitation model of re-redefining the urbanscape identity. This post-disaster spatial alteration was based on the continuity of the linear urban fabric of the city through the urban development in the western part of the city and continuity of the existing urban pattern on the surrounding slopes through the housing construction on the slopes, that appeared even before the Bosnian War, revealing the post-disaster urban rehabilitation model of spatial continuity. Due to the severe disaster impact and the extensive destruction, the post-disaster rehabilitation process of Sarajevo consists mostly of the city's physical reconstruction that is still in progress, revealing the post-disaster urban rehabilitation model of post-disaster reconstruction. As the result of the war, many open public and green spaces were transformed into cemeteries, which enhanced the spatial memorialization of the war, together with many monuments and memorials dedicated to the victims and spatial memory of the war, revealing the post-disaster urban rehabilitation model of memorialization (Table 8.1; Fig. 8.1).

8.1.2 The Battle of Grozny, (1994–1995)

The Pre-disaster Phase of the Disaster Process

The city of Grozny is located in the southwestern corner of the Chechen Republic of the Russian Federation. As the capital city of the Chechen Republic, the city was founded in 1818 and served as a military fortress.[13] During its urban development, the city became an industrial capital, consisting of two airports, 123 roads, and a mixture of multi-storey buildings, the main railroad line and the Sunzha River that divides the city into four parts.[14] The historical battle for Grozny began initially when Chechnya declared its independence from Russia on September, 61,991, which increased the tension and following a sequence of conflicts, resulted with the Russian armed intervention on December 11, 1994.[15]

The Disaster Event Phase of the Disaster Process

The battle of Grozny was a strategic operation plan aimed at separation of the city into zones, with the main strategy to attack the city from four main directions at the same time: the northern axis, the north-eastern axis, the western axis and the eastern axis. The initial impact started from the eastern axis, to first capture the bridges

[13] Thomas (2003)

[14] Thomas (2003:162)

[15] Thomas (2003:163)

Table 8.1 Disaster process overview of the Siege of Sarajevo

Disaster process overview	
Disaster category	Man-made
Disaster sub-category	Sociological disasters: war
Case study disaster (country, region/city, date, year of disaster event)	Bosnia and Herzegovina, Siege of Sarajevo, 1992–1996
Disaster process overview—use of the urbanscape	
The pre-disaster phase Prevention and mitigation	
Urban pattern of the everyday life	Urban vulnerability
The pre-disaster phase was mainly reflected through the peaceful urban environment provided through the safety and protection within the urbanscape Determining the factors for survival and evacuation during the disaster impact Identifying the appropriate roads for evacuation, construction of facilities that would provide rescue and shelter needs The longitudinal urban pattern of Sarajevo's spatial development represented a disadvantage in case of evacuation In case of war as a man-made disaster, the airport's location would represent a danger for its citizens since it is located in a densely populated urban environment within the city	Creating a plan of higher and lower zones of risk within the city; implementing self-supply stations in case of the need for water or electricity supply, with each zone accessible to two supply points; establishing shelters for people that would have to remain in the city during the disaster, by providing one shelter space for each citizen at home or at work Establishing shelters in the centre of the city, such as pedestrian tunnels, traffic roads, underground garages and buildings with dualistic purposes that would be used in the everyday life in the pre-disaster phase and during the disaster impact Since this study remained undeveloped, the lack of existing pre-disaster risk reduction plans and the necessary strategies and mitigation tools to cope with the disaster events, increased the vulnerability in case of a war even more

The disaster event phase Disaster impact—urbanscape identity disruption			
Direct disaster effect		Secondary effect	Indirect disaster effect
Population impact	10,514 civilians of which 1598 were children died or disappeared 1741 people of which 356 were children, were injured or permanently disabled During the war, 600,000 of the city's pre-war population left and 150,000 displaced population arrived	Disabling the function of the city's urban infrastructure network Disruption of the normal pattern of life and the everyday behaviour	Relocation of population Construction of massive and dense multi-storey apartment blocks in the western part of the city Urban sprawl on the surrounding slopes Alteration of the traditional landscapes Destruction of open public and green space and their transformation into cemeteries Re-defined urbanscape identity
Property impact	The extensive shell impact destroyed over 10,000 apartments and damaged 100,000 other buildings. A total of 23% of the property destruction was reported as seriously damaged, 64% as partially damaged and 10% as slightly damaged		

(continued)

Table 8.1 (continued)

The disaster event phase Disaster impact—urbanscape identity disruption			
Direct disaster effect		Secondary effect	Indirect disaster effect
Urban impact	Many green spaces and parks were transformed into cemeteries, and others have been deliberately contaminated by the aggressor with landmines. Due to the lack of heating supply, three-quarters of all urban trees and nearly all peri-urban trees within the siege line were cut down for firewood and household needs The main tram lines, buses and trolleys in the city were not able to function, only the Sarajevo airport, controlled by the United Nations Protection Force and the underground corridor underneath the airport, served as the main public transportation. The main tool for the disaster impact was used in the form of phone line, electricity, water and gas cut-offs. The river Miljacka was used for washing clothes and the waiting lines for getting water from alternative water sources were often under the impact of the surrounding snipers positioned on the hillsides		
Urban resilience	During the Siege, the surrounding hillside of Sarajevo's urbanscape was used as a strategic position in terms of the impact from the aggressor and at the same time as a defense line of resilience. Situated in the valley, the city favoured the defensive strategy in terms of creating a belt of urban resilience but at the same time, this layout enabled a more intense impact from		

(continued)

Table 8.1 (continued)

The disaster event phase Disaster impact—urbanscape identity disruption			
Direct disaster effect		Secondary effect	Indirect disaster effect
	the aggressor's side. The urban pattern of the every-day life shifted to another layer, below the city, in form of an underground corridor. The familiar movement within the existing urban structure influenced the survival routes, for water and food supply. For this reason, they were used as resilience routes, transformed to mis-guide the aggressor from the surrounding hillside, since they were continuously under the shell impact		
Identity impact	Destruction of cultural heritage and memory. One of these deliberate identity impacts was the destruction of the Bosnian national and university library, that resulted in the loss of 700 manuscripts and a unique collection of Bosnian publi-cations with the main aim to destroy the evidence of history		

The post-disaster phase Post-disaster recovery—post-disaster urbanscape identity rehabilitation		
Rehabilitation program	Rehabilitation plan	Urban rehabilitation strategies
Spatial Plan of Canton Sarajevo (1992–2015) Spatial Plan of Canton Sarajevo (2003–2023)	Post-disaster construction that is still in progress Possibility of reconcilia-tion with the past	Reconstruction of the urban infrastruc-ture Housing reconstruction Reconstruction of public buildings Reconstruction of religious, cultural and historical buildings Re-redefining the urbanscape identity Memorialization of the war (1992–1996) through monuments and memorials

The post-disaster urban rehabilitation models
Model of raising awareness for disaster risk and mitigation Model of urbanscape dualism Model of re-redefining the urbanscape identity Model of spatial continuity Model of post-disaster reconstruction Model of memorialization

Fig. 8.1 Photo of aerial view of Sarajevo in 1990, before the disaster impact; during the Siege (1992–1996); and present, during its post-disaster rehabilitation. (Authors, August 2022)

across the Sunzha River and to link with the northern and north-eastern forces, with the main aim to disrupt the central part of the city. The western impact followed the main rail line, with the main aim to occupy the train station and the presidential palace.[16] The plan was based on a "block by block" strategic approach.[17] With a defined strategic plan, the Russian armed forces attacked the city on December 31, 1994. Although the resistance of the Chechen forces was unpredictable, the city lacked a well-defined line of defence. The Russian forces attacked the city with an intensive shell impact of every 10 s over 3 h.[18] By January 10, they managed to make corridors to the city for the food supply and evacuation of the wounded. Although they managed to take over a larger part of the city by January 15, the Chechen forces still refused to leave the city easily.[19] By February, after almost 40 days of battle, the occupation was over with a final withdrawal of the Chechen forces from the city. The immense resistance of the Chechen forces to withstand the impact relied on their familiarity of the everyday life urban pattern, the use of the non-linear and asymmetric spatial lines, the use of mobile units rather than fixed defences and understanding the city at the micro level which did not rely only on the streets network and buildings as reference defensive points, but the use of other elements of the urban fabric instead.[20] During the battle, about 25,000 citizens died.[21] The city after the war was completely ruined; 90% of its central part and 70% of its suburban area were destroyed. The urban infrastructure experienced severe destruction of 100%.[22] Hundred thousands of people were forced to leave their homes and the population that remained in the small parts of the city was forced to survive without water, electricity and heating supply.

The Post-disaster Phase of the Disaster Process

In August 1996, the Chechen forces again reoccupied the city and forced the Russian forces to leave Chechnya, but in January 2000, during the Second Chechen-Russian conflict, the Russian forces occupied Grozny again.[23] Due to the forced resettlement, the pre-war population of Grozny was 401,000 and according to the records from 2001, about 183,000 people remained. Due to the severe destruction, the relocation of the population and extreme survival conditions, in 2006 the city established the rehabilitation program "The City Without Any Signs of War" for the revival of the city.[24] The main rehabilitation plan was based on the establishment of large-scale restoration and reconstruction projects. The project started with the debris removal within the city. By 2006, a total of 3216 properties, 50 educational and 51 health

[16] Thomas (2003:168)
[17] Thomas (2003:184)
[18] Thomas (2003:178)
[19] Thomas (2003:182)
[20] Thomas (2003:188)
[21] Rupe (1999:21)
[22] UN Habitat (2022:2)
[23] Thomas (2003:193)
[24] UN Habitat (2022:3)

facilities were completely restored and constructed. In terms of the urban infrastructure, more than 250 km of the city's road network was repaired, 13 bridges and 32 streets completely restored. The main aim of the post-disaster rehabilitation was to restore the pre-disaster way of life in order to ensure the return of the relocated population from their temporary shelters. During this period, 3773 families[25] that left during the disaster impact returned to the city. The post-war reconstruction was based on the transformation of the pre-disaster urban fabric of the city through major foreign investment projects. The post-disaster rehabilitation of the Chechen Republic is still in progress. As of October 2016, 53 investment projects have already been completed and 180 other investment projects are being implemented.[26] In Grozny, this is reflected through the design of architectural landmarks: such as the Akhmat Tower in 2014, to become a new symbol for the region. The construction of the tallest building in Chechnya and the second tallest in Russia is supposed to be completed in 2020[27] together with the construction of the Grozny Sea, Sports and Recreational Tourist Complex, the Grozny Mall, as a retail and entertainment area, and the Grozny Industrial Park for the industrial production and construction sector. Due to the severe destruction of the city, the city's urbanscape identity changed completely and led to the post-disaster urban reconstructing as a drastic revision of the city.[28]

The Post-disaster Urban Rehabilitation Models

The disaster process overview of the Battle in Grozny reveals several post-disaster urban rehabilitation models. The city's urbanscape was used by the Russian forces for a strategic approach for the impact, and at the same time it was used for the resistance of the Chechen forces within the city. Since the Russian forces used the urban division of the city to attack from different directions at the same time and the main urbanscape elements to capture the city with a block by block approach, the Chechen forces relied on their familiarity of the everyday life urban pattern, the use of the non-linear and asymmetric spatial lines and understanding the city at the micro level. The dualistic use of Grozny's urbanscape during the disaster impact and resilience reveals the post-disaster urban rehabilitation model of urbanscape dualism. The complete destruction of the city required Grozny to restore the pre-disaster way of life in order to ensure the return of the relocated population from their temporary shelters. The post-disaster urban rehabilitation of the city was based on large-scale restoration and reconstruction projects through major foreign investment projects, identifying the post-disaster urban rehabilitation model of post-disaster restoration and construction. This urbanscape transformation completely changed the city's

[25] UN Habitat (2022:4)

[26] PricewaterhouseCoopers (2017:16)

[27] Beardsley et al. (2018)

[28] Lunga (2011)

urbanscape identity, forming a new reconstructed urbanscape of new large-scale projects and architectural landmarks, revealing the post-disaster urban rehabilitation model of reconstructing the urbanscape identity. The post-disaster rehabilitation progress with the new urbanscape elements lacked the preservation of the memory of the Battle and its enhancement through the spatial memorialization, revealing the post-disaster urban rehabilitation model of disregarding the memory, which would erase the evidence of history and modify the identity (Table 8.2; Fig. 8.2).

8.1.3 The September 11 Terrorist Attacks in New York (2001)

The Pre-disaster Phase of the Disaster Process

The two Towers of the World Trade Centre represented the landmarks of the everyday urban life within the city of New York. This vertical commercial landscape of 110 storeys served as the main commercial and business centre in Manhattan, where more than 42,000 people worked.[29] The buildings were a physical constant and a confirmation of the New York's urbanscape, but they also reflected the changes within the city, exposing them as vulnerable to future disaster impacts. As such architectural and urban landmarks that dominated the New York skyline, they served as the main targets for the terrorist attack on September 11, 2001. This was not the first terrorist attack on the World Trade Centre. In 1993, an organized terrorist group detonated an extremely powerful bomb in the underground level of the building. The bomb caused damage to several floors of the underground garage and killed six people but left the towers undamaged.[30] After the bombing in 1993, many improvements were made in terms of emergency and evacuation measures to cope with future disaster impacts.

The Disaster Event Phase of the Disaster Process

On September 11, 2001, an organized terrorist group hijacked four California commercial airplanes with the main aim to carry out a coordinated attack on the United States. The first two airplanes hit the towers of the World Trade Centre. In the morning, at 8:45 a.m., the American Flight II, with 92 passengers and crew, crashed into the North Tower of the World Trade Centre. About 20 min later, the United Airlines Flight 175, with 65 passengers and the crew, hit the southern Tower of the World Trade Centre. At 9:40 a.m., the third airplane, the American Flight 77, with

[29] Sagalyn (2005:27)
[30] Fernandez (2002:10)

Table 8.2 Disaster process overview of the battle of Grozny

<table>
<tr><td colspan="4">Disaster process overview</td></tr>
<tr><td colspan="2">Disaster category</td><td colspan="2">Man-made</td></tr>
<tr><td colspan="2">Disaster sub-category</td><td colspan="2">Sociological disasters: war</td></tr>
<tr><td colspan="2">Case study disaster (country, region/city, date, year of disaster event)</td><td colspan="2">Chechen Republic of the Russian Federation, Battle of Grozny 1994–1995</td></tr>
<tr><td colspan="4">Disaster process overview—use of the urbanscape</td></tr>
<tr><td colspan="4">The pre-disaster phase
Prevention and mitigation</td></tr>
<tr><td colspan="2">Urban pattern of the everyday life</td><td colspan="2">Urban vulnerability</td></tr>
<tr><td colspan="2">The city was founded in 1818 and served as a military fortress that grew in an industrial capital, consisting of two airports, 123 roads, a mixture of multi-storey buildings, the main railroad line and the Sunzha River that divides the city into four parts
The historical battle for Grozny began initially when Chechnya declared its independence from Russia on September 6, 1991, which increased the tension and was followed by a sequence of conflicts, resulted in the Russian armed intervention on December 11, 1994</td><td colspan="2">The battle of Grozny was based on a strategic operation plan for the separation of the city into zones, with the main strategy to attack the city from four main directions at the same time: the northern axis, the north-eastern axis, the western axis and the eastern axis
The initial impact started from the eastern axis, to first capture the bridges across the Sunzha River and to link with the northern and north-eastern forces, with the main aim to disrupt the central part of the city. The wester impact followed the main rail line, with the main aim to occupy the train station and the presidential palace
The impact plan was based on a "block by block" strategic approach
The city lacked a well-defined line of defence</td></tr>
<tr><td colspan="4">The disaster event phase
Disaster impact—urbanscape identity disruption</td></tr>
<tr><td colspan="2">Direct disaster effect</td><td>Secondary effect</td><td>Indirect disaster effect</td></tr>
<tr><td>Population impact</td><td>About 25,000 citizens died</td><td rowspan="4">Hundred thousands of people were forced to leave their homes and the population that remained in the small parts of the city was forced to survive without water, electricity and heating supply</td><td rowspan="4">In August 1996, the Chechen forces again reoccupied the city and forced the Russian forced to leave Chechnya, but in January 2000, during the second Chechen–Russian conflict, the Russian forces again occupied Grozny
Due to the forced resettlement, the pre-war population of Grozny was 401,000. According to the records from 2001, about 183,000 people remained
Relocation of the population
New reconstructed urbanscape of new large-scale projects and architectural landmarks
Changed urbanscape identity</td></tr>
<tr><td>Property impact</td><td>The entire city was completely ruined; 90% of its central part was destroyed and 70% of its suburban area</td></tr>
<tr><td>Urban impact</td><td>The Russian forces attacked the city with an intensive shell impact of every 10 s over 3 h
The urban infrastructure was 100% destroyed</td></tr>
<tr><td>Urban resilience</td><td>The immense resistance of the Chechen forces relied on their familiarity of the everyday life urban pattern, the use of the non-linear and asymmetric spatial lines, the use of mobile units rather than fixed defences and understanding the city at</td></tr>
</table>

(continued)

Table 8.2 (continued)

The disaster event phase Disaster impact—urbanscape identity disruption			
Direct disaster effect		Secondary effect	Indirect disaster effect
	the micro level which didn't rely only on the streets network and buildings as reference defensive points, but the use of other elements of the urban fabric instead		
Identity impact	Identity modification through the neglect ion of the memory from the battle and without its preservation through spatial memorialization		

The post-disaster phase Post-disaster recovery—post-disaster urbanscape identity rehabilitation		
Rehabilitation program	Rehabilitation plan	Urban rehabilitation strategies
"The city without any signs of war" for the revival of the city	The establishment of large-scale restoration and reconstruction projects	To restore the pre-disaster way of life in order to ensure the return of the relocated population from their temporary shelters The project started with the debris removal within the city By 2006, a total of 3216 properties, of which 50 educational and 51 health facilities, were completely restored and constructed. In terms of the urban infrastructure, more than 250 km of the city's road network was repaired, 13 bridges and 32 streets completely restored The post-war reconstruction was based on the transformation of the pre-disaster urban fabric of the city, through major foreign investment projects Reconstructing the urbanscape identity Neglect ion of spatial memorialization of the Battle (1994–1995)

The post-disaster urban rehabilitation models
Model of urbanscape dualism Model of reconstructing the urbanscape identity Model of post-disaster restoration and construction Model of disregarding the memory

Fig. 8.2 Photo of aerial view of the centre of Grozny in the North Caucasian Federal District of Russia before the disaster; during the disaster impact; and during the post-disaster rehabilitation with the new presidential palace complex in the centre of Grozny, North Caucasian Federal District of Russia. (Authors, August 2022)

65 passengers and crew, struck into the Pentagon, and the fourth airplane, the United Airlines Flight 93, with 40 passengers and the crew that headed to Washington D.C. crashed into a field in Pennsylvania.[31] The towers resisted the aircraft impact for an hour, before collapsing to the ground. The aircraft impact caused massive structural damage to the floors and the inner core of the towers. In terms of the environmental impact, more than 1.2 million tons of debris material collapsed,[32] leaving a thick cloud of dust and causing the emission of toxic gases in the air that led to short-term health problems of the population that lived in the surrounding area.[33] During the building collapse, the amount of the gravitational energy converted into kinetic energy, radiating seismic waves that generated small earthquake shocks of the surrounding environment.[34] In total 2997 people died as the result from the initial impact: 2764 deaths were reported from the attack on the towers of the World Trade Centre, including the passengers from the two airplanes; 189 deaths were reported from the attack on the Pentagon, including the passengers from the airplane; and 44 deaths were reported from the airplane crash in Pennsylvania.[35] During the impact, the parks reaffirmed their meaning and values, as places of adjustment to cope with the disaster, but they also became important in the aftermath of the disaster, as places of healing and reflection.[36] For this reason, the visit of urban parks within the wounded urbanscape increased after the attack, but the visits to larger popular national parks decreased by more than 20%. The wounded urbanscape changed the interpretation of the public spaces and became a place of appreciation, where people went to "renew their spirit", seeking for security and peace.[37]

The Post-disaster Phase of the Disaster Process

The post-disaster recovery process was based on the rebuilding plan of the World Trade Centre, that included four key elements: restoring the historic streetscape; creating a new public transportation gateway; building an iconic complex of skyscrapers, instead of the seven buildings that were demolished on the site during the attack; and establishing the 9/11 Memorial.[38] The recovery process started with the debris removal of one million tons of steel and concrete, which was completed in May, 2002. The design for the new World Trade Centre complex of buildings was selected in 2003. The National September 11 Museum opened in May 2014 as a

---

[31] McDonnell (2004:10)

[32] Nordgrén et al. (2002:3)

[33] Nordgrén et al. (2002:5)

[34] Kausel (2002:15)

[35] START (2011:1)

[36] McDonnell (2004:90)

[37] McDonnell (2004:91)

[38] Sagalyn (2005:23)

symbol of resilience to preserve the memory and to commemorate with the victims, and the World Trade Centre Transportation Hub was officially opened in March 2016.[39] Throughout these last years, the construction activity for each design projects was gradually completed, but the redevelopment of the entire complex of the World Trade Centre is still in progress.

The Post-disaster Urban Rehabilitation Models

The disaster process overview of the terrorist attack on the World Trade Centre in New York reveals several post-disaster urban rehabilitation models. Because of the architectural and urban value of the two towers of the World Trade Centre, they dominated the New York urbanscape which made them the urban landmarks of the everyday life, but based on their urbanscape values, it also made them potential targets for future disaster impact. This dualistic urbanscape perception reveals the post-disaster urban rehabilitation model of urbanscape dualism. The use of parks and green spaces during the impact became extremely important for emergency evacuation and in the aftermath of the disaster, which reaffirmed the meaning and values of the urbanscape, revealing the need of the post-disaster urban rehabilitation model of reaffirming the urbanscape identity. The post-disaster recovery process consisted of rehabilitation strategies required to rebuild the complex of the World Trade Centre, which was reflected through the construction of different design projects, revealing the post-disaster urban rehabilitation model of rebuilding. The use of parks and green spaces in the aftermath of the disaster changed the interpretation of the public spaces as renewed places of appreciation in search for healing and reflection and the designing. The National September 11 Museum as a symbol of resilience reveals the need to preserve the memory and identifies the post-disaster urban rehabilitation model of memorialization. The need for preserving the memory of the tragic event and its enhancement through the physical interpretation of the past, by the rebuilt project on the same vulnerable site, reveals the post-disaster rehabilitation model of spatial continuity (Table 8.3; Fig. 8.3).

[39] Farrell (2017:67)

Table 8.3 Disaster process overview of the September 11 terrorist attacks in New York

Disaster process overview			
Disaster category		Man-made	
Disaster sub-category		Sociological disasters: terrorist attack	
Case study disaster (country, region/city, date, year of disaster event)		The United States of America, New York, World Trade Centre, September 11, 2001	
Disaster process overview—use of the urbanscape			
The pre-disaster phase Prevention and mitigation			
Urban pattern of the everyday life		Urban vulnerability	
The two towers of the World Trade Centre represented the landmarks of the everyday urban life within the city of New York. This vertical commercial landscape of 110 storeys served as the main commercial and business Centre in Manhattan.		The buildings were a physical constant and confirmation of the New York's urbanscape, but they also reflected the changes within the city, exposing them as vulnerable to future disaster impacts. As such architectural and urban landmarks that dominated the New York skyline, they served as the main targets for the terrorist attack on September 11, 2001	
The disaster event phase Disaster impact—urbanscape identity disruption			
Direct disaster effect		Secondary effect	Indirect disaster effect
Population impact	In total 2997 people died as the result from the initial impact: 2764 deaths were reported from the attack on the towers of the World Trade Centre, including the passengers from the two airplanes; 189 deaths were reported from the attack on the Pentagon, including the passengers from the airplane; and 44 deaths were reported from the airplane crash in Pennsylvania	During the building collapse, the amount of the gravitational energy converted into kinetic energy, radiating seismic waves that generated small earthquake shocks of the surrounding environment	Environmental impact Population health impact Changed interpretation of the public spaces as renewed places of appreciation
Property impact	The aircraft impact caused massive structural damage, to the floors and the inner core of the towers		
Urban impact	More than 1.2 million tons of debris material collapsed, leaving a thick dust and causing the emission of toxic gases in the air that led to short term health problems of the population that lived in the surrounding environment		

(continued)

Table 8.3 (continued)

The disaster event phase Disaster impact—urbanscape identity disruption			
Direct disaster effect		Secondary effect	Indirect disaster effect
Urban resilience	After the bombing in 1993, many improvements were made in terms of emergency and evacuation measures to cope with future disaster impacts The towers resisted the aircraft impact for an hour, before collapsing to the ground The use of parks and green spaces during the impact		
Identity impact	During the impact, the parks reaffirmed their meaning and values, as places of adjustment to cope with the disaster, but they also became important in the aftermath of the disaster, as places of healing and reflection		

The post-disaster phase Post-disaster recovery—post-disaster urbanscape identity rehabilitation		
Rehabilitation program	Rehabilitation plan	Urban rehabilitation strategies
Rebuilding the World Trade Centre complex	The restoration of the historic streetscape. Creating a new public transportation gateway. Building an iconic complex of skyscrapers, instead of the seven buildings that were demolished on the site, during the attack; and establishing the 9/11 memorial	The recovery process started with the debris removal of one million tons of steel and concrete, which was completed in May, 2002 The National September 11 Museum opened in May 2014 as a symbol of resilience to preserve the memory and to commemorate the victims, and the World Trade Centre Transportation Hub was officially opened in March 2016 The construction activity for each design projects was gradually completed, but the redevelopment of the entire complex of the World Trade Centre is still in progress

The post-disaster urban rehabilitation models
Model of urbanscape dualism Model of reaffirming the urbanscape identity Model of post-disaster rebuilding Model of memorialization Model of spatial continuity

Fig. 8.3 Photo of aerial view of the towers of the World Trade Centre before the attack, in March 2001; during the collapse on September 11, 2001; and the rebuilt complex of the World Trade Centre, in September 11, 2019. (Authors, August 2022)

References

Beardsley S, Stochetti A, Cerone M (2018) The design of Akhmat Tower. Int J Tall Build Urban Habitat 2:28–29

Englund GL (2015) Rebuilding Sarajevo: lessons for post-conflict urban reconstruction. The Allepo project. Center for conflict, negotiation and recovery. School of Public Policy. Central European University, Budapest

Ernst JZ, Vukićević B, Jakulj T, Ilich W (2003) Sarajevo paradox: survival throughout history and life after the Balkan war. InterMarium 6(3):1–15
Farrell M (2017) Fifteen years of rebuilding the World Trade Center. Cornell Real Estate Rev 15(1): 67. https://hdl.handle.net/1813/70789
Fernandez JE (2002) A brief history of the World Trade Center towers. In: Chapter 1: The towns lost and beyond. A collection of essays on the WTC by researchers at the Massachusetts Institute of Technology. Massachusetts Institute of Technology MIT, Cambridge, pp 5–12
Kausel E (2002) Inferno at the World Trade Center, NY. In: Chapter 2: The towns lost and beyond. A collection of essays on the WTC by researchers at the Massachusetts Institute of Technology. Massachusetts Institute of Technology MIT, Cambridge, pp 13–16
Lacan I, McBride JR (2009) War and trees: the destruction and replanting of the urban and peri-urban forest of Sarajevo, Bosnia and Herzegovina. Urban For Urban Green 8(3):133–148. https://doi.org/10.1016/j.ufug.2009.04.001
Lunga RD (2011) Urban models. In: Urban models and public–private partnership. Springer, Berlin, pp 7–46
Martín-Díaz J, Nofre J, Oliva M, Palma P (2015) Towards a sustainable urban development in post-war Sarajevo. Area 47(4):376–385. https://doi.org/10.1111/area.12175
McDonnell JA (2004) Responding to the September 11 terrorist attacks. National Center for Cultural Resources. U.S. Department of the Interior, Washington, DC
Nordgrén MD, Goldstein EA, Izeman MA (2002) The environmental impacts of the world trade center attacks. A preliminary assessment. Natural Resources Defense Council, NRDC, New York
Nurković R (2015) Contemporary aspects of spatial and urban planning in the Canton of Sarajevo. Eur J Geogr 6(1):23–36
Omićević N, Bojanić Obad Šćitaroci B (2018) The urbanscape duality during a disaster and post-disaster. Prostor 26:106–115. https://doi.org/10.31522/p.26.1(55).8
Pilav A (2012) Before the war, war, after the war: urban imageries for urban resilience. Int J Disaster Risk Sci 3:23–27. https://doi.org/10.1007/s13753-012-0004-4
PricewaterhouseCoopers (2017) Investor's guide to the Chechen Republic. PricewaterhouseCoopers, New York
Rupe CA (1999) The Battle of Grozny. Lessons for military operations on urbanized terrain. U.-S. Army Armor Center. Parameters 29(2):87–102. https://doi.org/10.55540/0031-1723.1935
Sagalyn L (2005) Chapter 2: The politics of planning the world's most visible urban redevelopment project, in contentious city. Russell Sage Foundation, New York, pp 23–72
START (2011) Background report: 9/11, ten years later. The National Consortium for the study of terrorism and responses to terrorism. University of Maryland, College Park
Thomas TL (2003) The 31 December 1994–8 February 1995 Battle for Grozny. In: Block by block: the challenges of urban operations. U.S. Army Command and General Staff College Press, Kansas
Troncota M (2015) Sarajevo—a Border City caught between it's multicultural past, the Bosnian war and a European future. Eur Secur 19:119–138. https://nbn-resolving.org/urn:nbn:de:0168-ssoar-464653
UN Habitat (2022) Grozny project. Post-conflict restoration of the City of Grozny program "the city without any sign of war". Application from the City of Grozny. UN Habitat, Nairobi
William FJ (1994) The battle of Sarajevo and the law of armed conflict. United Nations Security Council, New York

Chapter 9
Comparative Analysis of the Selected Technological Man-Made Disaster Process Case Studies

Abstract This chapter presents an analytical overview of the selected technological man-made disaster case studies in the world, caused by non-intentional human actions, that include technological or industrial disasters. Within the technological man-made disaster subcategory, the following disaster case studies are analysed: Deepwater Horizon Oil Spill in the Gulf of Mexico (2010); Covid-19 Pandemic (2019-Ongoing) and Beirut Port Explosion (2020), based on the use of the urbanscape, through its built and natural elements: the urban pattern of the everyday life and the urban vulnerability during the pre-disaster phase; the direct urban impact during the disaster event phase and the urban rehabilitation during the post-disaster phase.

Keywords Man-made disasters · Sociological disasters · War · Terrorism

9.1 Comparative Analysis of the Selected Technological Man-Made Disaster Process Case Studies

9.1.1 Deepwater Horizon Oil Spill in the Gulf of Mexico (2010)

The Pre-disaster Phase of the Disaster Process

On April 20, 2010, the drilling activities on the Deepwater Horizon mobile platform were finalized, after discovering several oil and gas zones at the Macondo Prospect.[1] The crew members were in preparation of a planned temporary abandonment procedure. This procedure included several activities to plug the well, in order to move the rig to a new drilling site and that a production installation could return to

[1] The Macondo Prospect (Mississippi Canyon Block 252, abbreviated MC252) is an oil and gas prospect in the United States Exclusive Economic Zone of the Gulf of Mexico, off the coast of Louisiana.

N. Omićević, B. Bojanić Obad Šćitaroci, *The Urban Rehabilitation of Post-Disaster Scapes*, Disaster Resilience and Green Growth,
https://doi.org/10.1007/978-981-19-9505-7_9

the Macondo Prospect and extract the hydrocarbons.[2] Due to engineering and human failures, the drilling unit exploded, caught fire and sank. The cement barrier intended to keep the hydrocarbons below the sea floor and prevent and mitigate that the blowout was not properly installed. The physical barriers, designed for disaster mitigation, were not properly constructed, tested or maintained.[3] In case of a blowout, the blowout preventer would activate, but due to the poor technical maintenance, this function failed.[4]

The Disaster Event Phase of the Disaster Process

On the evening of April 20, 2010, an explosion occurred on the Deepwater Horizon drilling rig, caught fire and after 2 days collapsed to the seabed at a depth of 1569 m. The explosion and fire caused the tragic death of 11 people, 17 people were injured and 119 crew members were rescued.[5] The explosion led to the rupture of the main pipe that was connecting the rig on the sea surface to the Macondo Prospect, which resulted in 87 days of massive and uncontrolled release of oil and natural gas discharge into the northern Gulf of Mexico. About 134 million gallons of oil were released into the ocean and at its maximum extent on June 19, the oil covered over 40,000 square kilometres of the ocean, leading to one of the biggest environmental disasters in history. The oil moved with the sea currents, winds and tides, contaminating more than 2100 km of shoreline, from eastern Texas to the Florida Panhandle,[6] directly affecting various coastal resources, finishing areas, wetlands and wildlife.[7]

The Post-disaster Phase of the Disaster Process

In the aftermath of the disaster, several rescue ships were sent to reduce the flames of the fire. Within a week after the explosion, several response techniques were undertaken to collect and remove the oil, by burning the surface oil and shoreline clean-up to reduce the effects of the disaster impact. The actions included dispersant application, freshwater releases and wildlife mitigation.[8] About seven million litres of chemical dispersants were used to break the oil on the sea surface into small droplets. Other activities included the physical removal and burning of oil which was floating on the water. The response activities on the shoreline were focused on the

---

[2] Makocha et al. (2019:65)
[3] Makocha et al. (2019:65)
[4] NOAA (2016:4)
[5] Hermilo (2012:104)
[6] NOAA (2016:2)
[7] Nadeau et al. (2014:1)
[8] NOAA (2016:8)

removal of the floating oil nearshore and the removal of oiled materials along the shoreline, with ongoing clean-up activities on the beaches, construction of berms to prevent the oil spreading as the relocation and rehabilitation of the wildlife.[9] The use of chemical dispersants, oil burning at the sea surface, synthetic-based drilling muds and the construction of berms to prevent the oil spreading on the shoreline, resulted in additional environmental consequences.[10]

The mitigation process in the aftermath of the disaster was conducted through five stages: the construction of a cofferdam, a riser insertion tube tool (RITT), the "top kill operation", a "top hat" and a capping stack.[11] The cofferdam was a dome placed on the leak to channel hydrocarbons with a pipe to a collection ship, but this effort failed, because it became clogged with hydrates. After the failure of the cofferdam, the top tube tool (RITT) was installed on the end of the broken riser to carry the oil and gas to the sea surface. During its placement from May 15 until May 25, the RITT managed to collect some of the discharged oil.[12] After the removal of the RITT, the "top kill" operation started, with the main aim to stop the oil and gas discharge, by pumping in synthetic-based drilling mud. After this operation, the "top hat" was placed, a dome which was smaller than the cofferdam, in order to channel the hydrocarbons to the ocean surface. During the period from June 3 until July 10, the "top hat" managed to collect nearly 15,000 barrels per day.[13] After the "top hat", a capping stack was installed, as a blowout prevention, which eventually closed on July 15.

As the largest environmental disaster in the U.S. history, the Deepwater Horizon oil spill led to discharge of 4.1 million barrels of oil into the Gulf of Mexico, the contamination of deep-ocean wildlife and more than 1600 km of shoreline, affecting multiple species of birds, fishes and sea turtles.[14]

The post-disaster rehabilitation process was based on the restoration and recovery of the Gulf of Mexico region. The recovery and restoration program consisted of three initiatives to address the incurred environmental damages[15]: The Natural Resource Damage Assessment (NRDA), with the main aim to retrieve the degraded natural resources and services; The National Fish and Wildlife Foundation (NFWF) settlement funding, with the main aim to reduce the risk of future disaster impact to the Gulf Coast, by creating and restoring barrier islands and implementing river diversion projects to preserve the coastal habitat; and the Restore act, with the main aim to address the environmental damages, by using the best available science that

---

[9] NOAA (2016:2)

[10] NOAA (2016:8)

[11] NOAA (2016:4)

[12] NOAA (2016:4)

[13] NOAA (2016:5)

[14] Makocha et al. (2019:67)

[15] ELI (2014:3)

would restore and protect the natural resources, ecosystems, coastal wetland and wildlife habitat.[16]

The Post-disaster Urban Rehabilitation Models

The disaster process overview of Deepwater Horizon oil spill in the Gulf of Mexico reveals several post-disaster urban rehabilitation models. Although disaster mitigation techniques existed, their poor technical maintenance and construction led to its failure and increased the exposure to the existing vulnerability, revealing the need of the post-disaster urban rehabilitation model of raising the technological disaster awareness. The variety of strategic response actions that were undertaken to reduce the dispersion of oil and gas during the disaster impact reveals the post-disaster urban rehabilitation model of resilience, to cope with the disaster impact. Due to the unprecedent use of chemical dispersants during the disaster mitigation process and evoking additional environmental damages, the research identifies the post-disaster urban rehabilitation model of continuity. The response activities that were undertaken on the shoreline included ongoing clean-up activities on the beaches and the relocation and rehabilitation of the wildlife, identifying the post-disaster urban rehabilitation model of coastal restoration. The post-disaster rehabilitation process was based on the restoration and recovery of the Gulf of Mexico region with the main aim to retrieve the degraded natural resources and to protect and preserve the ecosystem, revealing the post-disaster urban rehabilitation model of natural resource preservation (Table 9.1; Fig. 9.1).

9.1.2 COVID-19 Pandemic (2019—Ongoing)

The Pre-disaster Phase of the Disaster Process

In urban history, pandemics were the catalyst for shaping the city and contributing to its urban development. Being and open and resilient system that it always under the impact of change, every disruption can be perceived as an opportunity to understand the city and to reshape it.[17] The first catalyst in the urban history of pandemics, for reshaping the city, was the Black Death or Plague.[18] The fatal impact of the disease led to the development of epidemiological measures within the urban environment and to provide functional and safe urban spaces. The early attempts included the creation of parks and public squares, organized street networks, carrying out clean water supply and sewage system. The city with its dense urbanization generated the

[16] ELI (2014:4)

[17] Vaeceslav (2020:9)

[18] Caused by a bacterial infection, the Black Death spread around Europe between 1347 and 1351.

Table 9.1 Disaster process overview of the Deepwater Horizon oil spill in the Gulf of Mexico

Disaster process overview			
Disaster category	Man-made disaster		
Disaster sub-category	Technological disasters: transport accident		
Case study disaster (country, region/city, date, year of disaster event)	Deepwater Horizon oil spill, Gulf of Mexico, April 20, 2010		
Disaster process overview—use of the urbanscape			
The pre-disaster phase Prevention and mitigation			
Urban pattern of the everyday life	Urban vulnerability		
Drilling activities on the Deepwater Horizon mobile platform, after discovering several oil and gas zones at the Macondo Prospect The crew members were in preparation of a planned temporary abandonment procedure. This procedure included several activities to plug the well, in order to move the rig to a new drilling site and that a production installation could return to the Macondo Prospect and extract the hydrocarbons	The cement barrier, intended to keep the hydrocarbons below the sea floor and prevent and mitigate the blowout was not properly installed The physical barriers, designed for disaster mitigation, were not properly constructed, tested or maintained In case of a blowout, the blowout preventer would activate, but due to the poor technical maintenance, this function failed		
The disaster event phase Disaster impact—urbanscape identity disruption			
Direct disaster effect		Secondary effect	Indirect disaster effect
Population impact	The fire caused the tragic death of 11 people, 17 people were injured	The explosion led to the rupture of the main pipe that was connecting the rig on the sea surface to the Macondo Prospect, which resulted in 87 days of massive and uncontrolled release of oil and natural gas discharge into the northern Gulf of Mexico	The use of chemical dispersants, oil burning at the sea surface, synthetic-based drilling muds and the construction of berms to prevent the oil spreading on the shoreline, resulted in additional environmental consequences
Property impact	The drilling unit exploded, caught fire and sank		
Urban impact	About 134 million gallons of oil were released into the ocean and at its maximum extent on June 19, the oil covered over 40,000 square kilometres of the ocean The oil moved with the sea currents, winds and tides, contaminating more than 2100 km of shoreline, from eastern Texas to the Florida panhandle, directly affecting various coastal resources, finishing areas, wetlands and wildlife		

(continued)

Table 9.1 (continued)

The disaster event phase Disaster impact—urbanscape identity disruption			
Direct disaster effect		Secondary effect	Indirect disaster effect
Urban resilience	Several response techniques were undertaken to collect and remove the oil, by burning the surface oil and shoreline clean-up to reduce the effects of the disaster impact		

The post-disaster phase Post-disaster recovery—post-disaster urbanscape identity rehabilitation		
Rehabilitation program	Rehabilitation plan	Urban rehabilitation strategies
The restoration and recovery of the Gulf of Mexico region	The Natural Resource Damage Assessment (NRDA) The National Fish and Wildlife Foundation (NFWF) The restore act	To retrieve the degraded natural resources and services To reduce the risk of future disaster impact to the Gulf Coast, by creating and restoring barrier islands and implementing river diversion projects to preserve the coastal habitat To address the environmental damages, by using the best available science that would restore and protect the natural resources, ecosystems, coastal wetland and wildlife habitat
Model of raising technological disaster awareness Model of resilience Model of continuity Model of coastal restoration Model of natural resource preservation		

spread of the disease and at the same time became testing ground to cope with the impact of the contagious diseases.[19]

Although the world experienced several viral outbreaks in recent years, the potential outbreak of contagious diseases became a contested matter.[20] Due to the lack of regulations in the prevention and disaster management in case of the pandemics, the outbreak of the disease COVID-19 in 2019, revealed the urban vulnerability and the resilience of the cities worldwide.

The Disaster Event Phase of the Disaster Process

In December 2019, a respiratory disease was detected in the city of Wuhan, the capital of the Hubei Province in the People's Republic of China. The virus was

[19] Vaeceslav (2020:98)

[20] Ritchie and Gill (2021:3)

Fig. 9.1 Photo of the Deepwater Horizon drilling rig, before the disaster, during the disaster impact, on April 20, 2010, and the oil dispersion in the aftermath of the disaster. (Authors, August 2022)

identified as SARS-CoV-2, a severe respiratory system coronavirus, which can cause a fatal respiratory illness and was given the name as COVID-19. Since the mode of transmission and the nature of the virus were initially unknown, the virus muted and by January 2020, it infected over 96 million people and caused the death of two million people worldwide.[21] The emergence of the virus became a contested matter by scientists and medical officials. It is argued that the virus originated as a consequence of the human and animal interaction, derived from the transmission of a mutated bat and pangolin virus on the human population, while others consider that the virus was human induced in a laboratory. Due to its contested nature of origin, the virus is considered as a technological disaster of "organic origin or conveyed by biological vectors, which include the exposure to pathogenic micro-organisms, toxins and bioactive substances, outbreaks of epidemic diseases, plant or animal contagion and insect or other animal plagues, that may cause loss of life, illness and other health impacts, property damage, social and economic disruption and environmental damage".[22] The Global COVID-19 figures, obtained by the World Health Organization data, show that by July 2022, the virus affected 63 countries with 114 million confirmed cases and 2.3 million confirmed deaths.[23] In terms of the urban impact, the urban pattern of the everyday life changed and was forced to adapt to various measures in terms of restrictions in the use of open public spaces, introducing lockdown measures and mobility restrictions, ensuring physical distance between individuals and tightening hygiene measures with the necessity of wearing face masks, cancellation of joint activities and gatherings and shifting education and work to the "stay at home" concept.[24] These measures were not equally applied and not to the same extent in the countries affected. As a substitute for the use of the restricted areas, such as parks, roads and closed parking areas were transformed into areas for pedestrians and bikers and interior spaces were reorganized to replace the closed playgrounds. The open public spaces were marked with signalization boards in terms of their usage and the existing street network modified with road barricades. All these measures had a positive and a negative impact on the environmental quality and the natural resources, they contributed to the wildlife, improved the urban air quality and the quality of water resources, but at the same time increased the pollution, due to the large amount of sanitary disposal, used in the treatment of the disease and reduction in the recycling process. The "stay at home" concept increased the demand for housing in the rural settings and growing interest for rural life.[25]

The Post-disaster Phase of the Disaster Process

The COVID-19 pandemic raised many questions in terms of reshaping the urban environment and the use of the open public spaces, mostly because it is still ongoing and uncertain when it will end. The post-disaster phase of the ongoing pandemic referred to research recommendations for future urban management in order to give

[21] Martinez and Short (2021:3)

[22] UNISDR (2009:5)

[23] https://data.humdata.org/dataset/coronavirus-covid-19-cases-and-deaths

[24] Glavan et al. (2022:1)

[25] Glavan et al. (2022:4)

implications for post-COVID planning. The findings related to the improvement of the air and water resources quality can be used to identify the sources of the pollution and to take effective treatment measures of wastewater.[26] In order to apply effective distance measures in time of pandemics, streets need to be redesigned to accommodate the needs of the pedestrians and green spaces need to meet the outdoor activities of the citizens, in order to achieve urban integration of the city and the greenery and contribute to the pandemic resilience and climate adaptation.[27] Post-COVID research implies to rethink the understanding of urban resilience and sustainability within the urban environment. This concept refers to distribute diversity in the use of space and to avoid density, so that cities can become more centres of culture and entertainment and less centres for work and education. Since the COVID-19 pandemic revealed the vulnerability of the open public spaces, in terms of unequal accessibility, uneven distribution and occupation of insufficient area within the city, the advantages of using the open public spaces in case of a pandemic outbreak need to be more acknowledged and become subject to reactivation and enhancement.

The Post-disaster Urban Rehabilitation Models

The disaster process overview of the global COVID-19 pandemic reveals several post-disaster urban rehabilitation models. Although the world experiences several pandemic outbreaks in the last years, the lack of mitigation measures to cope with the impact of possible diseases revealed the need for the post-disaster rehabilitation model of raising the technological disaster awareness. The COVID-19 pandemic had a fatal population impact worldwide, but due to the epidemiological measures undertaken, it contributed to the improvement of the environmental quality and natural resources within the cities, revealing the post-disaster urban rehabilitation model of resilience. The use of open public spaces during the pandemic was firstly restricted for use, but in the aftermath of the disaster and still ongoing, the open public spaces became extremely important to cope with the disaster effects, revealing the post-disaster urban rehabilitation model of urbanscape dualism. Due to the uncertain end of the ongoing pandemic and its effects on the urban environment, the research reveals the post-disaster urban rehabilitation model of continuity. Since the future urban strategies of post-COVID planning reveal the need for reshaping the open public spaces within the city to be more acknowledged as places that can cope with possible pandemic outbreaks and integrate more greenery within the city to become more pandemic resilient, the research reveals the post-disaster urban rehabilitation model of urban regeneration (Table 9.2; Fig. 9.2).

---

[26] Sharifi and Khavarian-Garmsir (2020:4)

[27] Sharifi and Khavarian-Garmsir (2020:12)

Table 9.2 Disaster process overview of the COVID-19 pandemic

Disaster process overview			
Disaster category		Man-made disaster	
Disaster sub-category		Technological disasters: biological hazard	
Case study disaster (country, region/city, date, year of disaster event)		COVID-19 Pandemic, Worldwide, 2019-Ongoing	
Disaster process overview—use of the urbanscape			
The pre-disaster phase Prevention and mitigation			
Urban pattern of the everyday life		Urban vulnerability	
In urban history, pandemics were the catalyst for shaping the city and contributing to its urban development The city with its dense urbanization generated the spread of the disease and at the same time became testing ground to cope with the impact of the contagious diseases Although the world experienced several viral outbreaks in recent years, the potential outbreak of contagious diseases became a contested matter		Due to the lack of regulations in the prevention and disaster management in case of the pandemics, the outbreak of the disease COVID-19 in 2019, revealed the urban vulnerability and the resilience of the cities worldwide	
The disaster event phase Disaster impact—urbanscape identity disruption			
Direct disaster effect		Secondary effect	Indirect disaster effect
Population impact	Since the mode of transmission and the nature of the virus were initially unknown, the virus muted and by January 2020, it infected over 96 million people and caused the death of two million people worldwide The Global COVID-19 figures, obtained by the World Health Organization data, show that by July 2022, the virus affected 63 countries with 114 million confirmed cases and 2.3 million confirmed deaths	Various epidemiological measures such as restrictions in the use of open public spaces, introducing lockdown measures and mobility restrictions, ensuring physical distance between individuals and tightening hygiene measures with the necessity of wearing face masks, cancellation of joint activities and gatherings and shifting education and work to the "stay at home" concept	The epidemiological measures had a positive and a negative impact on the environmental quality and the natural resources, they contributed to the wildlife, improved the urban air quality and the quality of water resources, but at the same time increased the pollution, due to the large amount of sanitary disposal, used in the treatment of the disease and reduction in the recycling process. The "stay at home" concept increased the demand for housing in the rural settings and growing interest for rural life
Property impact	/		
Urban impact	The urban pattern of the everyday life changed and was forced to adapt to various epidemiological measures		

(continued)

Table 9.2 (continued)

The disaster event phase Disaster impact—urbanscape identity disruption			
Direct disaster effect		Secondary effect	Indirect disaster effect
Urban resilience	As a substitute for the use of the restricted areas, such as parks, roads and closed parking areas were transformed into areas for pedestrians and bikers and interior spaces were reorganized to replace the closed playgrounds. The open public spaces were marked with signalization boards in terms of their usage and the existing street network modified with road barricades		

The post-disaster phase Post-disaster recovery—post-disaster urbanscape identity rehabilitation		
Rehabilitation program	Rehabilitation plan	Urban rehabilitation strategies
The COVID-19 pandemic raised many questions in terms of reshaping the urban environment and the use of the open public spaces, mostly because it is still ongoing and uncertain when it will end	The post-disaster phase of the ongoing pandemic referred to research recommendations for future urban management in order to give implications for post-COVID planning	To identify the sources of the pollution and to take effective treatment measures of wastewater Streets need to be redesigned to accommodate the needs of the pedestrians and green spaces need to meet the outdoor activities of the citizens, in order to achieve urban integration of the city and the greenery and contribute to the pandemic resilience and climate adaptation To rethink the understanding of urban resilience and sustainability within the urban environment To distribute diversity in the use of space and to avoid density, so that cities can become more centres of culture and entertainment and less centres for work and education The advantages of using the open public spaces in case of

(continued)

Table 9.2 (continued)

The post-disaster phase Post-disaster recovery—post-disaster urbanscape identity rehabilitation		
Rehabilitation program	Rehabilitation plan	Urban rehabilitation strategies
		a pandemic outbreak need to be more acknowledged and become subject to reactivation and enhancement
The post-disaster urban rehabilitation models		
Model of raising technological disaster awareness Model of resilience Model of urbanscape dualism Model of continuity Model of urban regeneration		

9.1.3 Beirut Port Explosion (2020)

The Pre-disaster Phase of the Disaster Process

The Beirut Port was the country's largest port and the busiest in the Middle East, providing storage for most of Lebanon's grain reserves and as such played an important role in the country's economy. Situated in the city centre, the Port is in close proximity to the city's dense urban and historical core, commercial district and surrounded with the country's key private and state service, including medical infrastructures and educational facilities, the main electricity provider as well as the bus terminal.[28] The explosion occurred as Lebanon was already coping with the impact of the crisis in Syria, leading to the influx of 1.5 million refugees, since October 2019 the impact of the socio-economic crisis, leading to the increasing poverty and food insecurity and with the impact of the COVID-19 outbreak, since February 2020, leading to the disruption of the country's health system.[29] All these factors increased the city's vulnerability to cope with any possible disaster impact, but the maintenance and storage of a large concentration of ammonium nitrate were the main factor of vulnerability in the pre-disaster phase. Since 1916, the unproper use and storage of ammonium nitrate are responsible for the occurrence of 30 disasters.[30]

In 2013, the cargo ship "Rhosus" with 2750 tonnes of ammonium nitrate, docked at the Beirut Port. The ship left the Black Sea Port of Batumi, Georgia and was destined for Mozambique, but to the technical problems, the ship was forced to dock at Beirut's port on November 21, 2013. The ammonium nitrate cargo was transferred in early 2014 to Hangar 12 and it remained unproperly stored for 6 years, until the

[28] OCHA (2021:5)

[29] OCHA (2021:5)

[30] Tacchi (2021:5)

Fig. 9.2 Photo of the air pollution in New Delhi in November 2018, before the pandemic outbreak; a mass cremation of the died victims from the COVID-19 in New Delhi; and the improvement of air quality, due to the applied lockdown measures in April, 2020 in New Delhi. (Authors, August 2022)

disaster occurrence.[31] Regardless of the existing vulnerability, evacuation of the neighbourhoods in close proximity to the port was never initiated.

[31] Tacchi (2021:6)

The Disaster Event Phase of the Disaster Process

On August 4, 2020, around 17:45, a fire broke out in Hangar 12 warehouse at the Port of Beirut. Triggered by fireworks that were stored in the warehouse, around 18:07, the first explosion occurred and 30 s later a supersonic blast wave struck the entire city.[32] The blast created a 140 m wide crater at the Port[33] and was felt in Cyprus. It led to the death of 200 people, injured 6500 people, including approximately 1000 children,[34] caused an immense destruction of the city's urban fabric, by damaging at least six hospitals and 20 clinics, 163 schools and 9200 buildings within a 3 km radius from the Port, affecting 219,000 people and their displacement in the surrounding areas.[35] Despite existing timeframe from the fire outbreak and the explosion, no evacuation measures of the people living near the Port were undertaken.[36]

Alongside the population and property impact, the explosion destroyed the Port's silos that stored the country's grain reserves and the central warehouse of the Ministry of Public Health, which stored the medical supplies, vaccines and the personal protective equipment.[37] The explosion occurred whilst the country was struggling with the impact of the COVID-19 outbreak and its effects on the health system.[38]

In terms of the environmental impact, the explosion led to the dispersion of toxins and gases in the air and leaving a significant amount of debris.[39]

The Post-disaster Phase of the Disaster Process

In the aftermath of the explosion approximately 500 residents volunteered in the city's rehabilitation process and removal of the debris.[40] In order to gain the general overview of the city's destruction and to identify the most affected zones and buildings, the Beirut Municipality initiated a Rapid Building Level Damage Assessment.[41] The data from the assessment had to contribute to the prioritization and guidance of the suggested interventions of the most affected areas of Beirut.[42] As stated in the disaster event phase, the property impact affected 219,000 people, who were in need of shelter. The UN Refugee Agency repaired 1734 housing units and rehabilitated 1330 and supported the repair and rehabilitation of 123 heritage

---

32 Tacchi (2021:6)
33 Tacchi (2021:6)
34 OCHA (2021:5)
35 OCHA (2021:4)
36 Tacchi (2021:7)
37 OCHA (2021:5)
38 Tacchi (2021:15)
39 OCHA (2021:24)
40 Tacchi (2021:20)
41 Mansour and Sleiman (2020:2)
42 The World Bank (2020:9)

buildings.[43] In February 2022, the Town Planning and Heritage Sites Development Studio FZ-LLC, developed the project proposal "Beirut Port Reconstruction, An Integrated Participatory Urban Strategy", with the main aim to rehabilitate Beirut's Port, railway and northern coast. The proposed urban strategies refer to the strategic integration of the Railway and Port, a new Masterplan for the Port, memorialization of the remains of the Port silos and the explosion site to serve as a Memorial and landmark, land reclamation, environmental rehabilitation and coastal regeneration to preserve the coastal ecosystem and biodiversity.[44] In the same month, The Beirut Municipality and the Inspireli Awards[45] announced an unique architectural competition for students worldwide for the renewal of the Beirut Port. The aim of the competition was to envision an important redesign, but the crucial transport clusters in the post should be kept unchanged.[46] Although the memorialization of the remains of the Port silos would serve as urban regeneration in the post-disaster urban rehabilitation process, Lebanese officials have been warning about the possibility of the silos collapsing. On July 25, 2022, approaching the second Anniversary of the Blast, after a fire broke out, the silos partly collapsed. According to the Lebanese state media, in total four parts of the silos have fallen, and it is expected that the remainder of the silos will fall soon.[47]

The Post-disaster Urban Rehabilitation Models

The disaster process overview of the Beirut Port explosion reveals several post-disaster urban rehabilitation models. The unproper storage and maintenance of the ammonium nitrate in close proximity to the city centre reveal the post-disaster urban rehabilitation model of raising the technological disaster awareness. The immediate and effective response of the citizens in the aftermath of the disaster to volunteer in the removal of the debris and rehabilitation process reveals the post-disaster urban rehabilitation model of resilience. Due to the reparation of 1734 housing units housing units in the post-disaster phase of the disaster process, the research identifies the post-disaster urban rehabilitation model of reconstruction. Proposing the memorialization of the explosion site and the remains of the Port silos in terms of a memorial and landmark reveals the post-disaster urban rehabilitation model of memorialization. The post-disaster urban strategies refer to the environmental rehabilitation, coastal regeneration and preservation of the coastal ecosystem, revealing the post-disaster urban model of coastal restoration and the post-disaster urban rehabilitation model of natural resource preservation. With the aim to integrate the Beirut's Port and Railway, the research identifies the post-disaster urban rehabilitation model of urban regeneration (Table 9.3; Fig. 9.3).

---

[43] UNHCR (2021:4)

[44] Mrad and Khodor (2022)

[45] Global student contest in Architecture, Urban design and Landscape and Interior design

[46] https://bustler.net/news/8855/the-beirut-port-renewal-competition-shortlists-40-student-projects

[47] https://www.nytimes.com/2022/08/04/world/middleeast/beirut-port-silos-collapse-2020-explosion.html

Table 9.3 Disaster process overview of the Beirut Port explosion

Disaster process overview			
Disaster category		Man-made disaster	
Disaster sub-category		Technological disasters: industrial accident	
Case study disaster (country, region/city, date, year of disaster event)		Beirut Port Explosion, Lebanon, August 4, 2020	
Disaster process overview—use of the urbanscape			
The pre-disaster phase Prevention and mitigation			
Urban pattern of the everyday life		Urban vulnerability	
The Beirut port was the country's largest port and the busiest in the middle east, providing storage for most of Lebanon's grain reserves and as such played an important role in the country's economy. Situated in the city centre, the port is in close proximity to the city's dense urban and historical core, commercial districts and surrounded with the country's key private and state service, including medical infrastructures and educational facilities, the main electricity provider as well as the bus terminal		Lebanon was already coping with the impact of the crisis in Syria, leading to the influx of 1.5 million refugees, since October 2019 the impact of the socio-economic crisis, leading to the increasing poverty and food insecurity and with the impact of the COVID-19 outbreak, since February 2020, leading to the disruption of the country's health system The unproper maintenance of a large concentration of ammonium nitrate stored in Hangar 12 for 6 years at Beirut Port Evacuation of the neighbourhoods in close proximity to the port was never initiated	
The disaster event phase Disaster impact—urbanscape identity disruption			
Direct disaster effect		Secondary effect	Indirect disaster effect
Population impact	200 people died, 6500 people were injured, including approximately 1000 children	Triggered by fireworks that were stored in the warehouse, around 18:07, the first explosion occurred and 30 s later a supersonic blast wave struck the entire city	Despite existing timeframe from the fire outbreak and the explosion, no evacuation measures of the people living near the port were undertaken leading to a high number of injuries Displacement of the 219,000 affected people The combined struggle of the disrupted health system and the impact of COVID-19
Property impact	Damaging at least six hospitals and 20 clinics, 163 schools and 9200 buildings within a 3 km radius from the port, affecting 219,000 people The explosion destroyed the Port's silos that stored the country's grain reserves and the central warehouse of the Ministry of Public Health, which stored the medical supplies, vaccines and the personal protective equipment		

(continued)

Table 9.3 (continued)

<table>
<tr><td colspan="4">The disaster event phase
Disaster impact—urbanscape identity disruption</td></tr>
<tr><td colspan="2">Direct disaster effect</td><td>Secondary effect</td><td>Indirect disaster effect</td></tr>
<tr><td>Urban impact</td><td>An immense destruction of the city's urban fabric
The explosion led to the dispersion of toxins and gases in the air and leaving a significant amount of debris</td><td rowspan="2"></td><td rowspan="2"></td></tr>
<tr><td>Urban resilience</td><td>In the aftermath of the explosion approximately 500 residents volunteered in the city's rehabilitation process and removal of the debris</td></tr>
</table>

<table>
<tr><td colspan="3">The post-disaster phase
Post-disaster recovery—post-disaster urbanscape identity rehabilitation</td></tr>
<tr><td>Rehabilitation program</td><td>Rehabilitation plan</td><td>Urban rehabilitation strategies</td></tr>
<tr><td>Beirut municipality initiated a rapid building level damage assessment
Beirut port reconstruction, an integrated participatory urban strategy</td><td>To gain the general overview of the city's destruction and to identify the most affected zones and buildings
Rehabilitation of Beirut's port, railway and northern coast</td><td>The UN refugee agency repaired 1734 housing units and rehabilitated 1330 and supported the repair and rehabilitation of 123 heritage buildings
Beirut port reconstruction project with the main aim to rehabilitate Beirut's port, railway and northern coast
Architectural competition for students worldwide for the renewal of the Beirut port, with the aim to envision an important redesign, but the crucial transport clusters in the post should be kept unchanged
On July 25, 2022, approaching the second anniversary of the blast, after a fire broke out, the silos partly collapsed. According to the Lebanese state media, in total four parts of the silos have fallen, and it is expected that the remainder of the silos will fall soon</td></tr>
</table>

<table>
<tr><td>The post-disaster urban rehabilitation models</td></tr>
<tr><td>Model of raising technological disaster awareness
Model of resilience
Model of reconstruction
Model of memorialization
Model of coastal restoration
Model of natural resource preservation
Model of urban regeneration</td></tr>
</table>

Fig. 9.3 Photo of the Beirut Port before the disaster, during the explosions, on August 4, 2020, and in the aftermath of the disaster. (Authors, August 2022)

References

ELI (2014) Building bridges. Connecting the overlapping goals, resources and institutions of Gulf of Mexico. Restoration and conservation. Environmental Law Institute, ELI, Washington, DC

Glavan OL, Nikolić N, Folić B, Vitošević B, Mitrović A, Kosanović S (2022) Covid-19 and city space: impact and perspectives. Sustainability 14(3):1–19. https://doi.org/10.3390/su14031885

Hermilo RL (2012) The oil spill in the Gulf of Mexico. In: Experimental and theoretical advances in fluid dynamics. Springer, Berlin, pp 103–136. https://doi.org/10.1007/978-3-642-17958-7_8

https://bustler.net/news/8855/the-beirut-port-renewal-competition-shortlists-40-student-projects. Accessed 20 Aug 2022

https://www.nytimes.com/2022/08/04/world/middleeast/beirut-port-silos-collapse-2020-explosion.html. Accessed 15 Aug 2022

Makocha IR, Ete T, Saini G (2019) Deepwater Horizon oil spill: a review. Int J Tech Innov Mod Eng Sci 5(4):65–71

Mansour E, Sleiman A (2020) Beirut port explosions response. Beirut Municipality rapid building-level damage assessment. Municipality of Beirut and UN-Habitat, Lebanon

Martinez L, Short JR (2021) The pandemic city: urban issues in time of COVID-19. Sustainability 13(6):10. https://doi.org/10.3390/su13063295

Mrad EA, Khodor A (2022) Beirut port reconstruction: integrated participatory urban strategy—phase 1, project brief and planning directives report. Lebanon, TPHD Studio FZ-LLC

Nadeau L, Kaplan M, Sands M, Moore K, Goodhue C (2014) Assessing the impacts of the Deepwater Horizon oil spill on tourism in the Gulf of Mexico region. U.S. Department of the Interior Bureau of Ocean Energy Management, New Orleans

NOAA (2016) Chapter 2: Final programmatic damage assessment and restoration plan and final programmatic environmental impact statement. Incident overview. NOAA, Washington, DC, pp 1–22

OCHA (2021) The humanitarian response to Beirut port explosions Lebanon flash appeal—end report. OCHA, New York

Ritchie L, Gill D (2021) Considering COVID-19 through the lens of hazard and disaster research. Soc Sci 10(7):1–20. https://doi.org/10.3390/socsci10070248

Sharifi A, Khavarian-Garmsir AR (2020) The COVID-19 pandemic: impacts on cities and major lessons for urban planning, design, and management. Sci Total Environ 749:142391. https://doi.org/10.1016/j.scitotenv.2020.142391

Tacchi J (2021) An anatomy of the Beirut port blast. AOAV, London

The World Bank (2020) Beirut rapid damage and needs assessment. International bank for reconstruction and development. The World Bank, Washington, DC

UNHCR (2021) Beirut port explosions UNHCR's response update. The UN Refugee Agency, Lebanon

UNISDR (2009) 2009 UNISDR terminology on disaster risk reduction. UNISDR, Geneva

Vaeceslav M (2020) Post-pandemic city: historical context for new urban design. Transylv Rev Adm Sci 16:94–108. https://doi.org/10.24193/tras.SI2020.6

Chapter 10
Review of the Comparative Analysis of the Disaster Process Case Studies in the World

Abstract This chapter presents the results of the conducted analysis of the natural and man-made disaster case studies, based on the urban pattern of the everyday life and urban vulnerability, during the pre-disaster phase of the disaster prevention and mitigation process; the urban impact and urban resilience during the disaster event phase of the direct disaster impact and the urban rehabilitation programs, plans and urban rehabilitation strategies, implemented during the post-disaster recovery phase of the post-disaster rehabilitation process.

Keywords Urban vulnerability · Urban impact · Urban resilience · Urban rehabilitation

10.1 Review of the Comparative Analysis of the Disaster Process Case Studies in the World

The analytical overview of the disaster process of the selected case studies in the world was carried out based on the use of the urbanscape during the disaster process, through the urbanscape elements, to reveal the importance of the urbanscape during the disaster process, the urbanscape identity disruption and how the classification of the natural and man-made disasters influences the post-disaster rehabilitation of the disrupted urbanscape identity.

The comparative analysis of the urban pattern of the everyday life and urban vulnerability, during the pre-disaster phase of the disaster prevention and mitigation process, in the natural and man-made disasters, reveals how the use and the treatment of the open public spaces in the everyday life influence the resistance and the impact during a disaster. Their role in the pre-disaster phase of the process is influenced by their role of the everyday life that can increase or reduce the disaster impact through the existing vulnerability and influence the post-disaster recovery process. Therefore, the use of the urbanscape during the pre-disaster phase embodies a dualistic role that can enable mitigation and reduction during the disaster impact and at the same time because of its pre-disaster factors of vulnerability, increase the disaster impact. This dualistic role in natural disasters is reflected through the pre-disaster scale,

N. Omićević, B. Bojanić Obad Šćitaroci, *The Urban Rehabilitation of Post-Disaster Scapes*, Disaster Resilience and Green Growth,
https://doi.org/10.1007/978-981-19-9505-7_10

density and construction within an urban structure, that influences the limitation of the empty open public spaces, important for the evacuation, while in man-made disasters, this dualism is reflected through the urban pattern of the city's development, that influences the evacuation strategies, tools, routes and the location of strategic places of occupation in terms of sociological disaster impact. In the case of Sarajevo, the aggressor created an urban belt of occupation, considering the city's linear urban development, surrounded by the mountains. In the case of Grozny, the city was attacked from four main directions at the same time, considering the existing urban structure, where the main railroad line and the Sunzha River divide the city into four parts. The two towers of the World Trade Centre represented the city's landmark of the everyday life, which made them more exposed and vulnerable to become the main targets of the terrorist attack. The lack of existing pre-disaster risk reduction plans, the necessary strategies and mitigation tools to cope with the disaster events led to the lack of a well-defined line of defence in all sociological man-made disaster case studies.

In terms of technological disaster impact, the dualistic role in man-made disasters refers to unequal accessibility of open public spaces and their uneven distribution and occupation within the city, that influences the effectiveness of the epidemiological measures undertaken during pandemics and the use of mitigation tools to cope with the disaster effects. During the COVID-19 pandemic, city's worldwide became platforms for generating the disease and at the same time served as a testing ground to cope with the impact of the disease outbreak. The improper storage of ammonium nitrate in the close proximity of the city dense urban core, during the Beirut blast and the lack of maintenance of existing mitigation techniques during the Deepwater Horizon oil spill increased the exposure towards the existing vulnerability.

The lack of risk awareness and a failure of existing protection systems led to continuously increasing population density in the natural disaster-prone areas. Despite the existing vulnerability of the affected areas, they are still favoured for living, due to the available natural resources. The growing population density and increased mobility and urbanization resulted with land-use changes and land conversion in these hazard prone areas, such as deforestation of peatland and forestland conversion into agricultural land, increased coastal urbanization and modifying the natural soil of the slopes, by removing the existing vegetation and reducing the natural infiltration capacity of the natural soil. The comparative analysis of the natural case studies reveals a continuous and increasing exposure of the existing vulnerable areas to the future disaster impact.

The comparative analysis of the urban impact and urban resilience during the disaster event phase of the direct disaster impact, the secondary effects and the urbanscape identity disruption derived from indirect disaster effects, in the natural and man-made disasters, reveals the urbanscape duality, by promoting the role of security and vulnerability at the same time. In the examples of man-made disasters, the open public spaces within the urban fabric constitute challenging environments for impact operations and resistance, relying on the familiarity of the everyday use of the urban fabric and understanding the city at the micro level. During the Siege of Sarajevo, the surrounding hillside was used as a strategic position of attack and at the same time it was used as a defensive line of resistance. Being positioned in the valley, Sarajevo's urbanscape enabled a clear view for the aggressor's shell impact, but the familiar movement within the existing urban structure influenced the survival routes for water and food supply and the formation of urban barriers to misguide the sniper attacks from the surrounding hillside. During the COVID-19 outbreak, the open green spaces and parks were restricted to use. As a substitute for their usage, the roads and closed parking areas were transformed into areas for pedestrians and bikers and interior spaces were reorganized to replace the closed playgrounds, but in the aftermath of the disaster and still ongoing, the open public spaces became extremely important to cope with the disaster effects. The comparative analysis of the natural case studies shows how the familiar movement within the existing roads influenced the evacuation routes during the disaster impact. The unbuilt open spaces and parks enabled the possibility of establishing shelter zones, unplanned resettlement centres and urban barriers for resistance. The main difference between the natural and sociological man-made disaster urban impact refers to the man-made intentional nature of destruction. The comparative analysis of the man-made disaster urban impact consisted of extensive damage and destruction of the urban infrastructure, with the main aim to disable the city's transport and communication network. Damaging the essential infrastructure such as electricity and water supply disrupts the urban pattern of the everyday life. In the case of Sarajevo, the underground corridor was developed to serve as the main public transport, the river Miljacka was used for washing clothes, many green spaces and parks were deliberately contaminated with landmines and the urban trees were cut down for firewood and household. Disabling the water supply forced Sarajevo's citizens to search for other alternative forces. These waiting lines were often under the impact of the snipers from the surrounding hillside. The comparative analysis of the natural disaster urban impact shows extensive damage to the urban infrastructure and urban fabric, morphologic changes and widespread damage to protected land, forestry and agricultural land. Furthermore, the data from the comparative analysis of the natural and man-made disasters during the direct disaster effects in terms of the population and property impact is summarized in Table 10.1. In terms of the population impact in the natural case study sub-categories, the comparative analysis shows how the meteorological disasters have the highest death toll, as seen in Fig. 10.1. The comparison within the man-made disaster case study sub-categories shows how the biological technological disasters have the highest death toll, as seen in Fig. 10.2.

In terms of the property impact in the natural case study sub-categories, the comparative analysis shows how the meteorological disasters, beside the population impact, led the highest property destruction, followed by the hydrological disasters, as seen in Fig. 10.3. In the man-made disasters, the time frame of the disaster impact can last much longer than in natural disasters, which increases the severity of the destruction and damage more than in the natural disasters. The comparative analysis of the property impact in the man-made disaster case studies shows an immense property destruction in all case study sub-categories, except the biological technological disasters, who unlike its impact on the population, has no property impact, as seen in Fig. 10.4.

Table 10.1 Natural and man-made disaster impact, based on the population impact, through the death toll and the property impact, through the property destruction

Disaster sub-category		Case study disaster (country, region/city, date, year of disaster event)	Population impact (death toll)	Property impact (property destruction)
Hydrological disasters	Floods	United States of America, Gulf coast, August 23–31, 2005	1353	275,000
	Floods and landslides	Switzerland, August 21–23, 2005	6	/
	Floods	Pakistan, July, 18–September, 10, 2010	2000	728,192
Meteorological disasters	Cyclone	Bangladesh, April 29–30, 1991	140,000	1,000,000
	Cyclone	Queensland, Australia, March, 28–31, 2017	12	3300
	Hurricane	North-eastern coastal region, United States, March 29, 2012	162	650,000
Geophysical disasters	Volcanic eruption	Mount Pinatubo, Philippines, June 15, 1991	657	8260
	Earthquake and tsunami	Japan, Tohoku Region, March 11, 2011	15,879	125,530
	Earthquake and landslides	Nepal, Kathmandu Region, April 25, 2015	8891	605,254
Climatologic disasters	Wildfire	Indonesia, 1997	15,600	/
	Wildfire	United States of America, Pacific region/California, December, 2017 November, 2018	100	1063
	Heat wave and wildfire	Russia, June 21–August 19, 2010	62	3180

The man-made disaster impact—population and property impact

Disaster sub-category		**Case study disaster** (country, region/city, date, year of disaster event)	Population impact (death toll)	Property impact (property destruction)
Sociological disasters	War	Bosnia and Herzegovina, Siege of Sarajevo, 1992–1996	10,514	23%
	War	Chechen Republic of the Russian Federation, Battle of Grozny, 1994–1995	25,000	90%
	Terrorist attack	The United States of America, New York, World Trade Centre, September 11, 2001	2997	100%
Technological disasters	Transport accident	Deepwater Horizon Oil Spill, Gulf of Mexico, April 20, 2010	11	100%
	Biological hazard	COVID-19 Pandemic, 2019-Ongoing	2000,000	/
	Industrial accident	Beirut Port Explosion, Lebanon August 4, 2020	200	90%

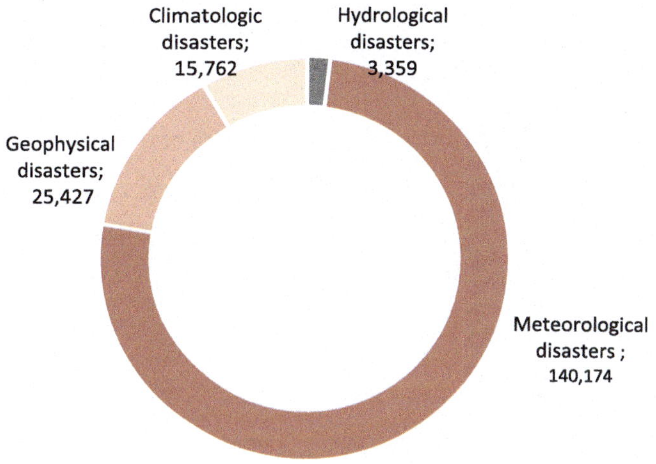

Fig. 10.1 The comparison of the death toll in the natural disaster case study sub-categories

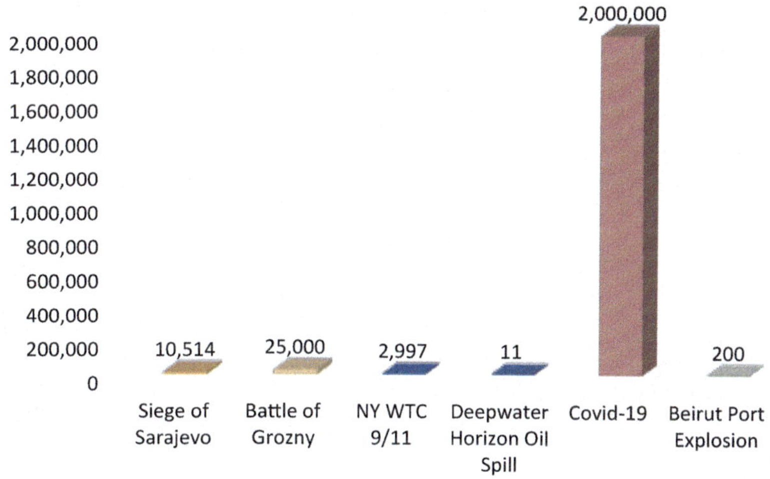

Fig. 10.2 The comparison of the death toll in the man-made disaster case study sub-categories

The secondary effects of the sociological man-made disaster impact disrupt the normal pattern of the everyday life much longer than in the natural disasters and technological man-made disasters, by disabling the main tools and strategies for survival. The secondary effects of the natural disaster impact refer to unprecedented scale of property, agricultural and forestry losses, landslides and debris flows in the cases of floods and dispersion of sulphur dioxide into the stratosphere, in the cases of volcanic eruptions, while the secondary effects of the technological man-made disaster impact refer to massive environmental impact and introducing restrictions in terms of the usage of open public spaces in case of pandemics.

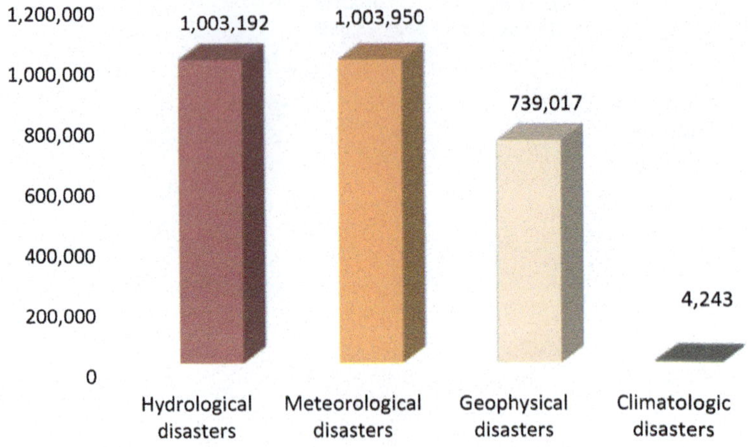

Fig. 10.3 The comparison of the property destruction in the natural disaster case study sub-categories

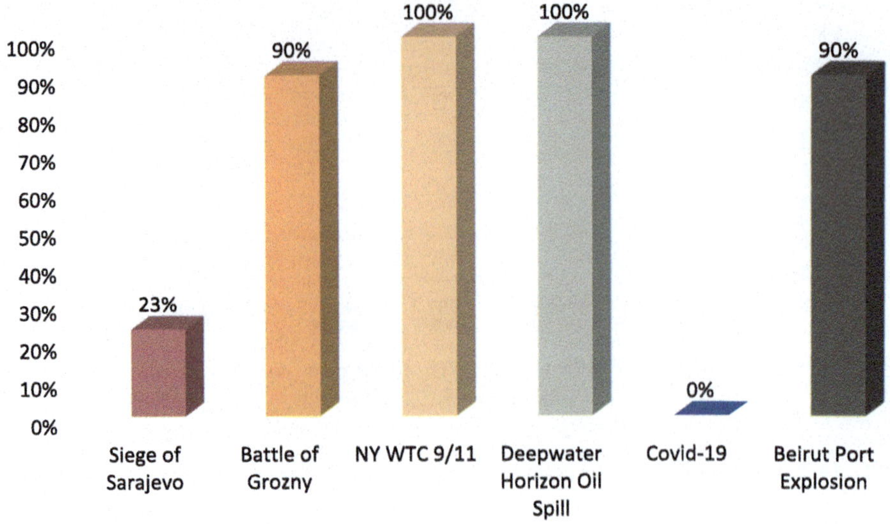

Fig. 10.4 The comparison of the property destruction in the man-made disaster case study sub-categories

The indirect effects of the natural disaster impact consist of temporary displacement of population and migration to new areas, alteration of natural landscape, by destroying endemic species of flora and fauna and destruction of forest and peatland, contamination of drinking water and long-term health effects leading to long-term evacuation of the people living in the affected area, such as the example of the nuclear disaster in the Fukushima Prefecture. Similar to the natural disaster impact, the indirect effects of the technological disaster impact refer to massive environmental consequences, displacement of population affected, but they also led to a positive and a negative impact on the environmental quality and natural resources, such as the improvement of air quality during the COVID-19 lock-down measures, but also the increased pollution, due to the large amount of sanitary disposal, used for the treatment of the disease.

The indirect effects of the sociological man-made disasters are influenced by the identity impact during the direct disaster effects. Unlike the natural disasters, sociological man-made disasters are provoked by intentional human actions through the deliberate impact on the urbanscape identity, heritage and memory, with the main aim to erase traces and evidence of history. In the case of Sarajevo, the indirect disaster effects refer to the redefined urbanscape identity based on the modification of Sarajevo's post-disaster urbanscape through the construction of massive and dense multi-storey apartment blocks in the western part of the city; alteration of the traditional landscape; destruction of open public spaces and green spaces and their transformation into cemeteries and the urban sprawl on the surrounding slopes. In the case of Grozny, the indirect effects refer to the reconstructed urbanscape identity, based on the reconstructed urbanscape of new large-scale projects and architectural landmarks. The reconstructed urbanscape identity erased the pre-disaster identity by neglecting the memory and failing to preserve the urban memorialization. In the case of the World Trade Centre terrorist attack, the indirect disaster effects refer to the reaffirmed urbanscape identity, based on the changed interpretation of the open public spaces as renewed places of appreciation. During the impact, the parks reaffirmed their meanings and values, as places of adjustment to cope with the disaster, but also as places of healing and reflection in the aftermath of the disaster.

The concept of duality within the urbanscape is also evident in the post-disaster recovery phase, reflected through the role of rehabilitation and through the opposing role derived from the indirect disaster effects. The scale and severity of the indirect effects from the disaster impact, in both natural and man-made disaster case studies, influence the efficacy and the implementation of the post-disaster rehabilitation process. This has been confirmed in the analysis of the sociological man-made disaster case studies, when the urbanscape identity becomes disrupted as the indirect disaster effect, derived from the initial disaster impact. The post-disaster urbanscape rehabilitation of the natural and technological man-made case studies consists of urban rehabilitation programs, plans and urban rehabilitation strategies which relate to developing mitigation measures to reduce the risk and vulnerability for future disaster impact and to recover and rebuild urban resilience of the hazard prone areas in order to raise the disaster risk awareness. The post-disaster urban rehabilitation of the sociological man-made disaster case studies is focused on the post-disaster reconstruction progress and transformation of the pre-disaster urbanscape, neglecting or accepting the memory, to restore the pre-disaster way of life by rehabilitating the urbanscape identity and the possibility of reconciliation with the past. Therefore, the research reveals how the rehabilitation of the disrupted urbanscape identity in the natural and technological man-made disaster case studies refers to its rebuilding and recovery, but in the man-made disasters case studies, the rehabilitation of the disrupted urbanscape identity refers to its re-redefinition, reconstruction or reaffirmation.

Chapter 11
Review of the Natural and Man-Made Disaster Process Case Studies Urban Rehabilitation Models

Abstract This chapter summarizes the existing urban rehabilitation models, derived from the conducted comparative analysis of the natural and man-made disaster process case studies. The research presents the existing criteria for urban planning, based on the urban rehabilitation plans and strategies used in the natural and man-made disaster case studies. With the unification of the existing urban planning criteria, derived from the urban rehabilitation models of the natural and man-made disaster process case studies, the research reveals the criteria for future urban planning.

Keywords Natural disasters · Man-made disasters · Urban rehabilitation · Urban planning

11.1 Review of the Natural and Man-Made Disaster Process Case Studies Urban Rehabilitation Models

The comparative analysis of the natural and man-made disaster process case studies disaster process shows how the post-disaster rehabilitation reveals several post-disaster urban rehabilitation models. These post-disaster urban rehabilitation models build upon the criteria of the urban vulnerability in the pre-disaster phase, the urban resilience during the disaster event phase and the urban rehabilitation strategies used in the post-disaster recovery process of both natural and man-made disasters. The natural disaster case studies rehabilitation models include the following models: the model of urban resilience, which combines the models of building resilience in case of floods, through controlled burning and natural regeneration in case of fire; the model of raising disaster awareness, which combines the models of raising disaster risk awareness in all natural disaster subcategories; the model of land-use relocation and readjustment refers to the combination of the urban plan redevelopment, utilization of land, relocation of agriculture lands, deforestation and land conversion; the model of disaster process mapping, which combines the models of developing geo-spatial maps to assess the vulnerability areas, and identifying possible shelter areas; the model of memorialization through monuments and memorials; the model

N. Omićević, B. Bojanić Obad Šćitaroci, *The Urban Rehabilitation of Post-Disaster Scapes*, Disaster Resilience and Green Growth,
https://doi.org/10.1007/978-981-19-9505-7_11

of adaptation to change which refers to the model of adaptation to the imposed changes; the model of natural recovery, which combines the models of reforestation and replanting, reducing the floods hazards, peatland restoration and conservation; the model of urban reconstruction which combines the models of redesign and reconstruction of the urban infrastructure and reuse of the debris material; the model of disaster impact continuity, which refers to the continuous exposure of the hazard prone countries to disaster; the model of disaster vulnerability assessment, which relates to the models of assessing the burned area in case of wildfires, and estimating the amount and volume of the debris flows in case of floods; and the model of population resettlement, which refers to the model of strategic population relocation and migration. The man-made case studies rehabilitation models include the following models: the model of raising disaster awareness, referring to the disaster risk and mitigation; the model of memorialization, through monuments and memorials; the model of urbanscape dualism, during the disaster process; the model of re-redefining the urbanscape identity, in the case of Sarajevo; the model of reconstructing the urbanscape identity, in the case of Grozny; the model of reaffirming the urbanscape identity, in the case of the World Trade Centre attack; the model of post-disaster reconstruction, which refers to the models of post-disaster reconstruction and restoration; the model of spatial continuity, referring to the post-disaster urban transformation; and the model of urban regeneration, referring to the reshaping of open public spaces and their functional integration within the city. The unification of the natural and man-made case studies urban rehabilitation models are summarized in Table 11.1.

11.1.1 Criteria for Future Urban Planning

Based on the comparative analysis of the natural and man-made disaster process case studies in the world, the research will reveal the existing criteria for urban planning, based on the urban rehabilitation plans and strategies used in the natural and man-made disaster case studies, to define the criteria for future urban planning. The system of the existing criteria for urban planning in the natural and man-made disaster process case studies is given in Table 11.2.

In order to define the criteria for future urban planning, it is important to classify the system of existing criteria for urban planning into criteria groups. With the unification of the existing urban planning criteria, derived from the urban rehabilitation models of the natural and man-made disaster process case studies summarized in Table 11.2, the research reveals the criteria for future urban planning, classified into five main groups: Criteria for environmental awareness; Criteria for natural preservation and management; Criteria for sustainable urbanscape development; Criteria for heritage and memory preservation and the Criteria for urbanscape re-affirmation. The system of criteria groups for future urban planning, based on the unification of the existing criteria for urban planning, is presented in Table 12.1.

Table 11.1 The natural and man-made disaster process case studies urban rehabilitation models

The natural disaster process case studies urban rehabilitation models						
Disaster sub-category		Case study disaster (country, region/city, date, year of disaster event)	Urban rehabilitation strategies	Post-disaster urban rehabilitation models	Unification of the natural case studies urban rehabilitation models	
Hydrological disasters	Floods	United States of America, gulf coast, August 23–31, 2005	Shifting the city's footprint and replacing the neighbourhood from the lower land level with green spaces Redevelopment of the neighbourhood plans for the flooded areas	Model of urban resilience Model of shrinking the city's footprint and land-use relocation Model of urban plan redevelopment Model of memorialization through monuments and memorials	Model of urban resilience Model of building coastal resilience Model of resilience by post-fire natural regeneration Model of urban resilience through controlled burning	Model of urban resilience
	Floods and landslides	Switzerland, August 21–23, 2005	Reducing the flood hazard, by increasing the drainage system, storage of runoff, enhancement of the river capacity and dams and flood embankments; reducing the flood volume and exposure, by flood plain regulations, redevelopment of land-use policies and housing and building codes; and preservation of the natural resources of floodplains, by enlarging the river space to enable space for extreme floods, soil pollution regulation and increasing the flood awareness	Model of urban resilience Model of reducing the flood hazard Model of reducing the flood volume and exposure Model of preserving the natural resources of the flood plain Model of raising the flood awareness	Model of raising the flood awareness Model of raising cyclone awareness Model of lack of awareness for the preservation of peatlands Model of raising awareness for wildfire mitigation	Model of raising disaster awareness
	Floods	Pakistan, July 18–September 10, 2010	The geo-spatial maps could conduct mapping in terms of vulnerability assessment of areas vulnerable to flood hazards, based on populations density and areas vulnerable to flood hazards to develop hazard management maps; identifying possible shelter sites in the pre-disaster phase	Model of mapping hazard prone areas Model of mapping shelter areas Model of mapping flood water diversion	Model of shrinking the city's footprint and land-use relocation Model of urban plan redevelopment Model of construction of	Model of land-use relocation and readjustment

(continued)

Table 11.1 (continued)

The natural disaster process case studies urban rehabilitation models						
Disaster sub-category		Case study disaster (country, region/city, date, year of disaster event)	Urban rehabilitation strategies	Post-disaster urban rehabilitation models	Unification of the natural case studies urban rehabilitation models	
			Possibility of utilizing flood water with the use of new channels for flood water diversion Assessment and analysis of the impact of flood in the agriculture areas Identifying the abandoned river channels to develop a multi-purpose drainage system, by activating the old drainage routes In 2015, development of the fourth national flood prevention plan (2015–2025)	Model of mapping abandoned river channels	multi-purpose cyclone shelters Model of utilizing newly accreted land Model of land-use relocation Model of land readjustment Model of relocation of agriculture land Model of deforestation and land conversion	
Meteorological disasters	Cyclone	Bangladesh, April 29–30, 1991	The mitigation measures involved the utilization of newly accreted land for planting mangrove plants, protection of the existing mangrove forest and replanting the degraded forests in order to reduce the water flow in case of storm surges	Model of urban resilience Model of construction of multi-purpose cyclone shelters Model of afforestation Model of utilizing newly accreted land Model of raising cyclone awareness	Model of mapping hazard prone areas Model of mapping shelter areas Model of mapping flood water diversion Model of mapping abandoned river channels	Model of disaster process mapping
	Cyclone	Queensland, Australia, March 28–31, 2017	In environmental terms, the plan encompasses several rehabilitation strategies: the coastal removal of debris, restoration of mangrove and other coastal vegetation and restoration of the beaches; clean-up and restoration of green waste; soil conservation; and improvement in	Model of raising cyclone awareness Model of urban resilience Model of coastal restoration	Model of memorialization through monuments and memorials	Model of memorialization

			the field of mapping and collecting data 8 months after the cyclone impact, all parks and forest have reopened or partially opened and all fire breaks have been repaired. The damaged roads have been reopened or provided alternatives. According to the transport reconstruction plan, all urban infrastructure reconstructions will be completed by June 2019			
	Hurricane	North-eastern coastal region, United States March 29, 2012	Housing reconstruction, restoration of the energy and transportation infrastructure, and developing coastline resilient projects, that include rebuilding beaches and board walls; and increasing coastal resilience to storm surges, by building up dunes and wetlands	Model of raising flood awareness Model of urban reconstruction Model of building coastal resilience Model of adaptation to change	Model of adaptation to change Model of adaptation to the imposed changes	Model of adaptation to change
Geophysical disasters	Volcanic eruption	Mount Pinatubo, Philippines June 15, 1991	The resettlement of the population groups, the Aeta highlanders and the displaced lowlanders Providing live hood programs focuses on the agriculture for the displaced farmers Providing basic social services and infrastructure Environmental management through the land use and replanning strategies	Model of population resettlement Model of resuming the pre-disaster way of life Model of adaptation to the imposed changes Model of continuous disaster impact	Model of reducing the flood hazard Model of reducing the flood volume and exposure Model of preserving the natural resources of the flood plain Model of afforestation Model of coastal restoration Model of reforestation Model of natural recovery Model of reforestation and replanting Model of peatland restoration and conservation	Model of natural recovery

(continued)

Table 11.1 (continued)

The natural disaster process case studies urban rehabilitation models

Disaster sub-category		Case study disaster (country, region/city, date, year of disaster event)	Urban rehabilitation strategies	Post-disaster urban rehabilitation models	Unification of the natural case studies urban rehabilitation models	
	Earthquake and tsunami	Japan, Tohoku region, March 11, 2011	Relocation of land use. Raising the base elevation of land and the height of the flood bank Reconstruction of housing property Redevelopment of urban infrastructure and facilities	Model of land-use relocation Model of land readjustment Model of urban reconstruction	Model of urban reconstruction Model of reuse of debris material Model of the disposal of the non-usable debris material Model of redesign of road networks	Model of urban reconstruction
	Earthquake and landslides	Nepal, Kathmandu region, April 25, 2015	The development of spatial planning in terms of reduction of disaster risks for the resettlement of the affected population and the relocation of agriculture and infrastructure of the affected areas, in the aftermath of the disaster The reuse and recycle of the debris material The management for the disposal of the non-usable debris material The redesign and strategic reconstruction of the road network, especially in the rural areas The improvement of water resources in case of water contamination Integrating farms in forest land to restore the agriculture sector and reforestation, by replanting the trees and vegetation in the disaster affected areas Over 130,000 buildings have been mapped based on their building type and construction characteristics, in order to achieve valuable data for future response and disaster risk reduction.	Model of relocation of agriculture land Model of reuse of debris material Model of the disposal of the non-usable debris material Model of redesign of road networks Model of reforestation Model of strategic population resettlement Model of urban resilience	Model of continuous disaster impact Model of continuity	Model of disaster impact continuity

Climatologic disasters	Wildfire	Indonesia, 1997	The availability of seeds for the development of the post-fire secondary forest The importance of natural regeneration through mixed planting of native species; and the awareness of the environmental importance of the secondary forests	Model of lack of awareness for the preservation of peatlands Model of deforestation and land conversion Model of continuity Model of resilience by post-fire natural regeneration Model of natural recovery	Model of vulnerability assessment Model of assessment of the burned area Model of estimating debris flows	Model of disaster vulnerability assessment
	Wildfire	United States of America, Pacific region/California, December, 2017; November, 2018	Assessment of the burned area Examining the possibility of evoking post-fire hazards, such as debris flows, based on the probability model and the volume model	Model of urban resilience through controlled burning Model of continuity Model of vulnerability assessment Model of assessment of the burned area Model of estimating debris flows	Model of strategic population resettlement	Model of population resettlement
	Heat wave and wildfire	Russia, June 21–August 19, 2010	The establishment of protected areas within the forest and replanting larches The restoration of more than 40,00 ha od degraded peatland and the development of guidelines for the peatland conservation and rehabilitation	Model of continuity Model of raising awareness for wildfire mitigation Model of natural recovery Model of reforestation and replanting Model of peatland restoration and conservation		

(continued)

Table 11.1 (continued)

The man-made disaster process case studies urban rehabilitation models

Disaster sub-category		Case study disaster (country, region/city, date, year of disaster event)	Urban rehabilitation Strategies	Post-disaster urban rehabilitation models	Unification of the man-made case studies urban rehabilitation models	
Sociological disasters	War	Bosnia and Herzegovina, Siege of Sarajevo, 1992–1996	Reconstruction of the urban infrastructure Housing reconstruction Reconstruction of public buildings Reconstruction of religious, cultural and historical buildings Re-redefining the urbanscape identity Memorialization of the War (1992–1996) through monuments and memorials	Model of raising awareness for disaster risk and mitigation Model of urbanscape dualism Model of re-redefining the urbanscape identity Model of spatial continuity Model of post-disaster reconstruction Model of memorialization	Model of raising awareness for disaster risk and mitigation	Model of raising disaster awareness
					Model of memorialization	Model of memorialization
					Model of urbanscape dualism	Model of urbanscape dualism
	War	Chechnya, Battle of Grozny 1994–1995	To restore the pre-disaster way of live in order to ensure the return of the relocated population from their temporary shelters The project started with the debris removal within the city By 2006, 3216 properties, 50 educational and 51 health facilities were completely restored and constructed. In terms of the urban infrastructure, more than 250 km of the city's road network was repaired, 13 bridges and 32 streets completely restored The post-war reconstruction was based on the transformation of the pre-disaster urban fabric of the city, through major foreign investment projects	Model of urbanscape dualism Model of reconstructing the urbanscape identity Model of post-disaster restoration and construction Model of disregarding the memory	Model of re-redefining the urbanscape identity	Model of re-redefining the urbanscape identity
					Model of reconstructing the urbanscape identity	Model of reconstructing the urbanscape identity

			Reconstructing the urbanscape identity Neglection of spatial memorialization of the Battle (1994–1995)			
	Terrorist attack	United States of America, World Trade Centre, New York, September 11, 2001	The recovery process started with the debris removal of one million tons of steel and concrete, which was completed in May, 2002 The National September 11 Museum opened in May 2014 as a symbol of resilience to preserve the memory and to commemorate with the victims and the World Trade Centre Transportation Hub was officially opened in March 2016 The construction activity for each design projects was gradually completed, but the redevelopment of the entire complex of the World Trade Centre is still in progress	Model of urbanscape dualism Model of reaffirming the urbanscape identity Model of post-disaster rebuilding Model of memorialization Model of spatial continuity	Model of reaffirming the urbanscape identity	Model of reaffirming the urbanscape identity
Technological disaster	Transport accident	Deepwater horizon oil spill, Gulf of Mexico, April 20, 2010	To retrieve the degraded natural resources and services To reduce the risk of future disaster impact to the Gulf Coast, by creating and restoring barrier islands and implementing river diversion projects to preserve the coastal habitat To address the environmental damages, by using the best available science that would restore and protect the natural resources, ecosystems, coastal wetland and wildlife habitat	Model of raising technological disaster awareness Model of resilience Model of continuity Model of coastal restoration Model of natural resource preservation	Model of post-disaster reconstruction Model of post-disaster restoration and construction Model of post-disaster rebuilding.	Model of post-disaster reconstruction
	Biological hazard	COVID-19 Pandemic, 2019-Ongoing	To identify the sources of the pollution and to take effective treatment measures of wastewater Streets need to be redesigned to accommodate the needs of the pedestrians and green spaces need to meet the outdoor activities of the citizens, in order to achieve urban integration of the city and the greenery and contribute to the pandemic resilience and climate adaptation	Model of raising technological disaster awareness Model of resilience Model of urbanscape dualism Model of continuity	Model of spatial continuity	Model of spatial continuity

(continued)

Table 11.1 (continued)

The man-made disaster process case studies urban rehabilitation models

Disaster sub-category		Case study disaster (country, region/city, date, year of disaster event)	Urban rehabilitation Strategies	Post-disaster urban rehabilitation models	Unification of the man-made case studies urban rehabilitation models	
			To rethink the understanding of urban resilience and sustainability within the urban environment To distribute diversity in the use of space and to avoid density, so that cities can become more centres of culture and entertainment and less centres for work and education The advantages of using the open public spaces in case of a pandemic outbreak need to be more acknowledged and become subject to reactivation and enhancement.	Model of urban regeneration		
	Industrial accident	Beirut port explosion, Gulf of Mexico, August 4, 2020	The UN refugee agency repaired 1734 housing units and rehabilitated 1330 and supported the repair and rehabilitation of 123 heritage buildings Beirut port reconstruction project with the main aim to rehabilitate Beirut's port, railway and northern coast Architectural competition for students worldwide for the renewal of the Beirut port, with the aim to envision an important redesign, but the crucial transport clusters in the post should be kept unchanged On July 25, 2022, approaching the second anniversary of the blast, after a fire broke out, the silos partly collapsed. According to the Lebanese state media, in total four parts of the silos have fallen, and it is expected that the remainder of the silos will fall soon	Model of raising technological disaster awareness Model of resilience Model of reconstruction Model of memorialization Model of coastal restoration Model of natural resource preservation Model of urban regeneration	Model of urban regeneration	Model of urban regeneration

					Model of coastal restoration Model of natural resource preservation	Model of natural recovery
					Model of resilience	Model of urban resilience

Table 11.2 The system of criteria groups for future urban planning based on the unification of the existing criteria for urban planning

	Criteria for future urban planning			
Unification of the natural and man-made case studies urban rehabilitation models		Existing criteria for urban planning	Criteria for future urban planning	Criteria group for future urban planning
Natural case studies urban rehabilitation models	Man-made case studies urban rehabilitation models			
Model of urban resilience	Model of raising disaster awareness	Criteria for raising disaster awareness Model of raising disaster awareness	Criteria for raising disaster awareness	Criteria for environmental awareness
Model of raising disaster awareness	Model of memorialization	Criteria for natural preservation and recovery Model of land-use relocation and readjustment Model of natural recovery	Criteria for reduction of disaster risk	
Model of land-use relocation and readjustment	Model of urbanscape dualism	Criteria for reduction of disaster risk Model of disaster process mapping Model of disaster vulnerability assessment Model of population resettlement	Criteria for developing disaster resilience	
Model of disaster process mapping	Model of re-redefining the urbanscape identity	Criteria for developing disaster resilience Model of adaptation to change Model of urban resilience	Criteria for reduction of disaster risk	

		Model of re-redefining the urbanscape identity Model of reaffirming the urbanscape identity Model of reconstructing the urbanscape identity		
Model of memorialization	Model of reconstructing the urbanscape identity	Criteria for memory preservation Model of memorialization	Criteria for natural preservation and recovery	Criteria for natural preservation and management
Model of adaptation to change	Model of reaffirming the urbanscape identity	Criteria for post-disaster urban construction projects Model of urban regeneration Model of urban reconstruction Model of post-disaster reconstruction	Criteria for post-disaster urban construction projects	Criteria for sustainable urbanscape development
Model of natural recovery	Model of post-disaster reconstruction	Criteria for spatial continuity Model of disaster impact continuity Model of spatial continuity	Criteria for memory preservation	Criteria for heritage and memory preservation
Model of urban reconstruction	*Model of spatial continuity*	Criteria for urbanscape dualism awareness Model of urbanscape dualism	Criteria for spatial *continuity*	Criteria for *urbanscape* Reaffirmation
Model of disaster impact continuity;	Model of urban regeneration		Criteria for urbanscape dualism awareness	
Model of disaster vulnerability assessment	Model of natural recovery			
Model of population resettlement	Model of urban resilience			

According to the unification of the criteria for future urban planning and their unification in six main groups, the research will elaborate every criteria group, based on the conducted analysis of the existing criteria from the natural and man-made disaster case studies.

11.1.2 The Comparative Analysis of the Criteria Group for Future Urban Planning

The Criteria Group for Environmental Awareness

This criteria group include the criteria for raising disaster awareness, the disaster vulnerability assessment, for developing disaster resilience and for reduction of disaster risk.

Based on the review of the comparative analysis of the disaster process case studies in the world, the research has shown how the lack of existing pre-disaster risk reduction plans, strategies and mitigation tools to cope with the disaster events influenced the lack of a defined defence line in all man-made disaster case studies. The lack of risk awareness and the failure of the existing protection system led to the increasing population density in the natural disaster-prone areas. This resulted with land-use changes and land conversion in the hazard prone areas, such as deforestation of peatland, forest land conversion into agricultural land and modifying the natural soil of the slopes.

These criteria involve mitigation measures to decrease the risk and vulnerability for future disaster impact and to recover and rebuild hazard prone areas in order to raise the disaster risk awareness. The mitigation measures consist of developing land-use planning policies, adaptation of the use of the buildings during the disaster impact and the development of geo-spatial maps, by mapping and collecting data for vulnerability assessment and possible shelter areas to develop disaster prevention plans.

The Criteria Group for Natural Preservation and Management

This criteria group include the criteria for preservation of natural resources and recovery.

The review of the comparative analysis of the disaster process case studies in the world shows morphologic changes and widespread damage to protected land, forestry and agricultural land. This resulted with an unprecedented scale of agricultural and forestry losses, landslides and debris flows in the case of floods leading to temporary displacement of population or migration to new areas and alteration of natural landscape, by destroying endemic species of flora and fauna and destruction of forest and peat land, contamination of drinking water and long-term health effects.

These criteria involve the development of urban design and land-use strategies and the preservation of natural resources, such as enlarging the river spaces in flooded areas, regulation of soil pollution, afforestation in the hazard prone areas and replanting degraded forests. They involve strategies in terms of land-use and re-planning strategies, reuse and recycle of debris material, regeneration of natural vegetation and restoration of the natural ecosystem, examination of areas evoking geomorphologic hazards, defining the protected green areas and implementing the adaptation of natural elements within the urban planning.

The Criteria Group for Sustainable Urbanscape Development

This criteria group include the criteria for post-disaster urban construction projects.

Based on the review of the comparative analysis of the disaster process case studies in the world, the research shows how the main difference between the natural and man-made disaster urban impact refers to the man-made intentional nature of destruction. This consisted of extensive damage and destruction of urban infrastructure, with the main aim to disable the city's transport and communication network. Damaging the essential infrastructure such as electricity and water supply disrupts the urban pattern of the everyday life, just like during the Siege of Sarajevo.

These criteria relate to the post-disaster restoration and reconstruction of housing, transport, urban reconstruction, the development of new-large-scale foreign investment projects as architectural landmarks and the process of introducing new urban concepts, connected to the traditional ones and enhancing the city's cultural identity through the new urban development process. In terms of sustainability, they involve planning measures that include mapping the road networks and boundaries, school, health facilities and buildings, based on their type and construction characteristics.

The Criteria Group for Heritage and Memory Preservation

This criteria group include the criteria for preservation of memory.

The review of the comparative analysis of the disaster process case studies in the world shows how the man-made disaster, unlike the natural disasters are provoked by deliberate impact on the urbanscape identity, through the destruction of heritage and memory, as evidence of history. The research reveals how the rehabilitation of the disrupted urbanscape identity in the natural case studies refers to its rebuilding and recovery, but in the man-made disaster case studies, the rehabilitation of the disrupted urbanscape identity refers to its re-redefinition, reconstruction or re-affirmation. After the War, in the case of Sarajevo and the World Trade Centre, the built environment became a stage of memorialization, through the streets, squares and bridges with commemorative signs as tools of visual remembrance. In the case of Grozny, the new reconstructed urbanscape lacked the preservation of memory, with the aim to erase history and modify the identity.

These criteria relate to the reconciliation process, by reconstructing the demolished religious, cultural and historical buildings, memorialization through monuments and memorials as symbols of resilience to preserve memory.

The Criteria Group for Urbanscape Reaffirmation

This criteria group include the criteria for spatial continuity and urbanscape dualism awareness.

Based on the review of the comparative analysis of the disaster process case studies in the world, the research shows how the use of the urbanscape embodies a dualistic role during the entire disaster process, in the pre-disaster phase, the urbanscape enables mitigation and reduction and at the same time increases the disaster impact, due to the existing pre-disaster vulnerability. During the disaster impact, the urbanscape duality is reflected by promoting security and vulnerability, in the form of attack and resistance and at the same time, as survival. In the post-disaster event phase, this dualistic role is evident through the post-disaster recovery process and the opposing role of (trans)formation, derived from the indirect disaster effects and the source of healing and remembrance, enforcing continuity and sustainability of the city's re-redefined urbanscape identity.

These criteria relate to the reaffirmation of the meaning and value of the open public spaces, before, during and after the disaster process. They are reflected through their dualistic role of security and vulnerability in the pre-disaster phase, impact and resistance and survival, during the disaster phase and through healing and remembrance in the post-disaster phase of the disaster process.

Chapter 12
Conclusion on the Main Findings

The contribution of the conducted research is reflected through the extensive analysis of the use of the urbanscape during the entire disaster process. The research first set the criteria for the analysis of the chosen disaster case studies, based on their category and subcategory division and the period of their occurrences, and then based on the conducted review of the existing research, defined the phases that comprise the disaster process overview. The research subsequently explained each disaster phase and the elements it comprises, in order to conduct the comparative analysis of the chosen disaster case studies in the world, based on the use of the urbanscape elements during each disaster phase and the interconnectedness of its elements between the disaster process, with the aim to reveal the importance of the urbanscape network in every phase of the disaster process.

The analytical overview of the disaster process case studies in the world was based on the use of the urbanscape, through its built and natural environment, reflected through the urban pattern of the everyday life and the urban vulnerability during the pre-disaster phase; the urban impact and urban resilience during the disaster event phase and the urban rehabilitation strategies used during the post-disaster phase.

The analytical overview of the disaster process case studies in the world was conducted to reveal the importance of the use of the urbanscape during the disaster process, through its concept of duality, by promoting the role of security and vulnerability at the same time; the urbanscape identity disruption, in terms of its change during the post-disaster phase of the disaster process and how the classification of the natural and man-made disasters influences the post-disaster rehabilitation of the disrupted urbanscape identity, concluding that the rehabilitation of the disrupted urbanscape identity in the natural case studies, refers to its rebuilding or recovery, but in man-made disaster case studies, the rehabilitation of the disrupted urbanscape identity refers to its redefinition, reconstruction or reaffirmation.

The research on the disaster process overview was conducted by observing the natural disaster aspect, sociological as intentional man-made disaster aspect and the

N. Omićević, B. Bojanić Obad Šćitaroci, *The Urban Rehabilitation of Post-Disaster Scapes*, Disaster Resilience and Green Growth,
https://doi.org/10.1007/978-981-19-9505-7_12

technological, as unintentional man-made disaster aspect, where both natural and sociological aspects intertwine. The main purpose of the results presented was to create awareness that the natural, sociological and technological disaster aspect should not be observed through their effects. In terms of future urban planning, it is important to understand that all three disaster aspects have the same impact on the urbanscape, as presented in Table 12.1, with the assumption that the urban impact of the natural, sociological and technological disasters can be modified. In all three disaster aspects, these urbanscape elements are related to: **Morphologic changes and widespread damage destruction of the natural environment through protected land, coastal, forestry and agricultural land; immense destruction of the city's urban fabric and urban infrastructure; and transformation of the urban pattern of the everyday life.** The modification of these urbanscape elements can be achieved through a preparatory planning strategy, which will mitigate future disaster risk and exposure and direct how to overcome them. Based on these results, the research shows that, despite the nature of the disasters, the preparatory planning mitigation of the natural, sociological and technological disasters should be approached in the same manner, which confirm the originality of this study and shows an important contribution to the research.

Table 12.1 The urbanscape elements affected during the natural, sociological and technological disaster impact

Disaster sub-category		Case study disaster (country, region/city, date, year of disaster event)	Urban impact Impact on the urbanscape elements during the disaster event phase of the disaster process	Urban impact of the natural, sociological and technological disasters
Hydrological disasters	Floods	United States of America, Gulf coast, August 23–31, 2005	The hurricane caused modification of coastal areas, as the result of the flood, high waves, waterborne contamination, and severe erosion The main communication facilities and vehicles were damaged or completely destroyed The existing protection system broke in over 50 locations. The levees and the floodwalls broke, resulting in flooding of 80% of the urban fabric	Morphologic changes and widespread damage destruction of the natural environment through protected land, coastal, forestry and agricultural land Immense destruction of the city's urban fabric and urban infrastructure Transformation of the urban pattern of the everyday life
	Floods and landslides		The floods enhanced many	

(continued)

Table 12.1 (continued)

Disaster sub-category		Case study disaster (country, region/city, date, year of disaster event)	Urban impact Impact on the urbanscape elements during the disaster event phase of the disaster process	Urban impact of the natural, sociological and technological disasters
		Switzerland, August 21–23, 2005	morphologic changes: the bank slopes collapsed; the erosion on the river caused the shifting of the channels; dike failure and overflows; and upwelling of ground water	
	Floods	Pakistan, July, 18–September, 10, 2010	About 20% of the country's total area was underwater. The devastating rain caused the Indus River to breach levees and overflow into flooding in the northwest and east Pakistan, affecting both rural and urban population The damage to the urban infrastructure and facilities was extensive	
Meteorological disasters	Cyclone	Bangladesh, April 29–30, 1991	The coastal embankments could not resist the high storm surge and the areas without sufficient afforestation got easily damaged The destruction of urban infrastructure was widespread. Many bridges got washed away and the transportation infrastructure was damaged	Morphologic changes and widespread damage destruction of the natural environment through protected land, coastal, forestry and agricultural land Immense destruction of the city's urban fabric and urban infrastructure Transformation of the urban pattern of the everyday life
	Cyclone		Roads, railways and bridges were	

(continued)

Table 12.1 (continued)

Disaster sub-category		Case study disaster (country, region/city, date, year of disaster event)	Urban impact Impact on the urbanscape elements during the disaster event phase of the disaster process	Urban impact of the natural, sociological and technological disasters
		Queensland, Australia, March, 28–31, 2017	damaged or blocked by fallen trees Out of 1040 natural preserved areas in Queensland, 223 areas of protected land were affected by the floods A total of 223 parks and forests within the Queensland coast were affected by the cyclone	
	Hurricane	North-eastern coastal region, United States, March 29, 2012	The storm flooded coastal roadways, tunnels, leaving more than half of the nation's daily transport system without service The storm surge caused dune and beach erosion, island breaching and inundation of wetlands, increasing the contamination of drinking water	
Geophysical disasters	Volcanic eruption	Mount Pinatubo, Philippines, June 15, 1991	Around 19,799 hectares of plantations, 10,206 hectares of agro-forestry land and 96,200 hectares of agricultural land were destroyed The impact led to a widespread damage to crops and livestock and urban infrastructure	Morphologic changes and widespread damage destruction of the natural environment through protected land, coastal, forestry and agricultural land Immense destruction of the city's urban fabric and

(continued)

Table 12.1 (continued)

Disaster sub-category		Case study disaster (country, region/city, date, year of disaster event)	Urban impact Impact on the urbanscape elements during the disaster event phase of the disaster process	Urban impact of the natural, sociological and technological disasters
	Earthquake and Tsunami	Japan, Tohoku region, March 11, 2011	Ruining 23,600 hectares of farm-land and altering the coastline with the 0.6 m subsidence of 400 km of Japan's east coast Damage to many road facilities, especially the main route along the coast, damages to bridge bearings and joints, the water drainage systems, electrical power station and electricity transmission facilities Many bridge decks and road embankments got washed away	urban infrastructure Transformation of the urban pattern of the everyday life
	Earthquake and landslides	Nepal, Kathmandu Region, April 25, 2015	Immense damage to the agriculture land, forest and water resources. About 2.2% of the forest land within the affected areas was lost	
Climatologic disasters	Wildfire	Indonesia, 1997	During the fire, up to ten million ha of forest and non-forest land was affected in Indonesia	Morphologic changes and widespread damage destruction of the natural environment through protected land, coastal, forestry and agricultural land
	Wildfire	United States of America, Pacific region/California, December, 2017; November, 2018	The fire burned the supporting vegetation of the rocks that protect the slope, changed the soil structure and	

(continued)

Table 12.1 (continued)

Disaster sub-category		Case study disaster (country, region/city, date, year of disaster event)	Urban impact Impact on the urbanscape elements during the disaster event phase of the disaster process	Urban impact of the natural, sociological and technological disasters
			compromised the stability of the landscape The infrastructure impact can be caused indirectly, since the landscape alteration can increase the post-disaster vulnerability with the appearance of debris flows and debris slides that could damage the transport and communication facilities within this vulnerable area	
	Heat wave and wildfire	Russia, June 21–August 19, 2010	The fires devastated the forest and crop land. About 30,376 forest fires were registered, including 1162 peat fires. About five million ha of forest land was burned	
Disaster sub-category		Case study disaster (country, region/city, date, year of disaster event)	Urban impact—Impact on the urbanscape elements during the disaster event phase of the disaster process	Urban impact of the natural, sociological and technological disasters
Sociological disasters	War	Bosnia and Herzegovina, Siege of Sarajevo, 1992–1996	Many green spaces and parks were transformed into cemeteries and others have been deliberately contaminated by the aggressor with	Morphologic changes and widespread damage destruction of the natural environment through protected land, coastal, forestry

(continued)

Table 12.1 (continued)

Disaster sub-category		Case study disaster (country, region/city, date, year of disaster event)	Urban impact—Impact on the urbanscape elements during the disaster event phase of the disaster process	Urban impact of the natural, sociological and technological disasters
			landmines. Due to the lack of heating supply, three-quarters of all urban trees and nearly all peri-urban trees within the siege line were cut down for firewood and household needs The main tram lines, buses and trolleys in the city were not able to function, only the Sarajevo airport, controlled by the United Nations Protection Force and the under-ground corridor underneath the air-port, served as the main public trans-portation. The main tool for the disaster impact was used in the form of phone line, electricity, water and gas cut offs. The river Miljacka was used for washing clothes and the waiting lines for getting water from alterna-tive water sources were often under the impact of the surrounding snipers positioned on the hillsides	and agricultural land Immense destruc-tion of the city's urban fabric and urban infrastruc-ture Transformation of the urban pattern of the everyday life
	War	Chechen Repub-lic of the Russian Federation,	The Russian forces attacked the city with an intensive	

(continued)

Table 12.1 (continued)

Disaster sub-category		Case study disaster (country, region/city, date, year of disaster event)	Urban impact—Impact on the urbanscape elements during the disaster event phase of the disaster process	Urban impact of the natural, sociological and technological disasters
		Battle of Grozny, 1994–1995	shell impact of every 10 s over 3 h The urban infrastructure was 100% destroyed	
	Terrorist attack	The United States of America, New York, World Trade Centre, September 11, 2001	More than 1.2 million tons of debris material collapsed, leaving a thick dust and causing the emission of toxic gases in the air that led to short term health problems of the population that lived in the surrounding environment	
Technological disasters	Transport accident	Deepwater Horizon Oil Spill, Gulf of Mexico, April 20, 2010	About 134 million gallons of oil were released into the ocean and at its maximum extent on June 19, the oil covered over 40,000 square kilometres of the ocean The oil moved with the sea currents, winds and tides, contaminating more than 2100 km of shoreline, from eastern Texas to the Florida Panhandle, directly affecting various coastal resources, finishing areas, wetlands and wildlife	Morphologic changes and widespread damage destruction of the natural environment through protected land, coastal, forestry and agricultural land Immense destruction of the city's urban fabric and urban infrastructure Transformation of the urban pattern of the everyday life
	Biological hazard	COVID-19 Pandemic, 2019-Ongoing	The urban pattern of the everyday life changed and	

(continued)

Table 12.1 (continued)

Disaster sub-category		Case study disaster (country, region/city, date, year of disaster event)	Urban impact—Impact on the urbanscape elements during the disaster event phase of the disaster process	Urban impact of the natural, sociological and technological disasters
			was forced to adapt to various epidemiological measures	
	Industrial accident	Beirut Port Explosion, Lebanon August 4, 2020	An immense destruction of the city's urban fabric The explosion led to the dispersion of toxins and gases in the air and leaving a significant amount of debris	

SPRINGER NATURE

GPSR Compliance

The European Union's (EU) General Product Safety Regulation (GPSR) is a set of rules that requires consumer products to be safe and our obligations to ensure this.

If you have any concerns about our products, you can contact us on ProductSafety@springernature.com

In case Publisher is established outside the EU, the EU authorized representative is:

Springer Nature Customer Service Center GmbH
Europaplatz 3
69115 Heidelberg, Germany

The manufacturer's authorised representative in the EU is Springer Nature Customer Service Centre GmbH, Europaplatz 3, 69115 Heidelberg, Germany. If you have any concerns regarding our products, please contact ProductSafety@springernature.com

Printed and bound by CPI Group (UK) Ltd, Croydon, CR0 4YY
01/10/2026
02225882-0007